T0330545

TOTAL
ENGINEERING
QUALITY
MANAGEMENT

QUALITY AND RELIABILITY

A Series Edited by

EDWARD G. SCHILLING

Coordinating Editor
Center for Quality and Applied Statistics
Rochester Institute of Technology
Rochester, New York

W. GROVER BARNARD
Associate Editor for
Human Factors
Vita Mix Corporation
Cleveland, Ohio

RICHARD S. BINGHAM, JR.
Associate Editor for
Quality Management
Consultant
Brooksville, Florida

LARRY RABINOWITZ
Associate Editor for
Statistical Methods
College of William and Mary
Williamsburg, Virginia

THOMAS WITT
Associate Editor for
Statistical Quality Control
Rochester Institute of Technology
Rochester, New York

TOTAL
ENGINEERING
QUALITY
MANAGEMENT

RONALD J. COTTMAN

Consultant
Fountain Hills, Arizona

ASQC Quality Press **Milwaukee**

Marcel Dekker, Inc. **New York • Basel • Hong Kong**

Library of Congress Cataloging-in-Publication Data

Cottman, Ronald J.
    Total engineering quality management / Ronald J. Cottman.
        p.   cm. -- (Quality and reliability ; [37])
    Includes bibliographical references and index.
    ISBN 0-8247-8740-4 (alk. paper)
    1. Production management--Quality control.   2. Total quality
management.   I. Title.   II. Series.
    TS156.C6167   1992
    658.5'62--dc20   92-25552
                                                                  CIP

This book is printed on acid-free paper.

ASQC
Quality Press
611 East Wisconsin Avenue
Milwaukee, Wisconsin 53202

Marcel Dekker, Inc.
270 Madison Avenue
New York, New York 10016

Current printing (last digit):

10 9 8 7 6 5 4 3 2 1

PRINTED IN THE UNITED STATES OF AMERICA

ABOUT THE SERIES

The genesis of modern methods of quality and reliability will be found in a simple memo dated May 16, 1924, in which Walter A. Shewhart proposed the control chart for the analysis of inspection data. This led to a broadening of the concept of inspection from emphasis on detection and correction of defective material to control of quality through analysis and prevention of quality problems. Subsequent concern for product performance in the hands of the user stimulated development of the systems and techniques of reliability. Emphasis on the consumer as the ultimate judge of quality serves as the catalyst to bring about the integration of the methodology of quality with that of reliability. Thus, the innovations that came out of the control chart spawned a philosophy of control of quality and reliability that has come to include not only the methodology of the statistical sciences and engineering, but also the use of appropriate management methods together with various motivational procedures in a concerted effort dedicated to quality improvement.

This series is intended to provide a vehicle to foster interaction of the elements of the modern approach to quality, including statistical applications, quality and reliability engineering, management, and motivational aspects. It is a forum in which the subject matter of these various areas can be brought together to allow for effective integration of appropriate techniques. This will promote the true benefit of each, which can be achieved only through their interaction. In this sense, the whole of quality and reliability is greater than the sum of its parts, as each element augments the others.

The contributors to this series have been encouraged to discuss fundamental concepts as well as methodology, technology, and procedures at the leading edge of the discipline. Thus, new concepts are placed in proper perspective in these evolving disciplines. The series is intended for those in manufacturing, engineering, and marketing and management, as well as the consuming public, all of whom have an interest and stake in the improvement and maintenance of quality and reliability in the products and services that are the lifeblood of the economic system.

The modern approach to quality and reliability concerns excellence: excellence when the product is designed, excellence when the product is made, excellence as the product is used, and excellence throughout its lifetime. But excellence does not result without effort, and products and services of superior quality and reliability require an appropriate combination of statistical, engineering, management, and motivational effort. This effort can be directed for maximum benefit only in light of timely knowledge of approaches and methods that have been developed and are available in these areas of expertise. Within the volumes of this series, the reader will find the means to create, control, correct, and improve quality and reliability in ways that are cost effective, that enhance productivity, and that create a motivational atmosphere that is harmonious and constructive. It is dedicated to that end and to the readers whose study of quality and reliability will lead to greater understanding of their products, their processes, their workplaces, and themselves.

Edward G. Schilling

PREFACE

A quality product, as well as quality *as* a product, begins at the most basic level. A quality product begins with the individual worker and requires innovative thinking. In the past, Quality Control (QC)—and in some cases, Quality Assurance (QA)—was seen as an inspection device to be focused on engineering errors. Under today's management concepts, this is no longer acceptable. Quality Assurance must be managed so that every person in every department understands that quality begins in individual application toward a total quality operation. Total Quality Management (TQM) is just that: the management of quality during the total length of a project or process. Total Engineering Quality Management results in total engineering quality.

Engineering project managers interested in entering the world of TQM find themselves in a peculiar position. They can pick up any catalog of publications on the subject and see a multitude of available literature; however, most of the writing in this book deals with TQM in the manufacturing or production environment. When engineering project managers can find material dealing with TQM within the engineering environment, then the volume includes no hint of how to measure progress; Statistical Process Control, the definitive method for measuring quality progress, is left for another day and another book; and metrics are hardly discussed. The reader finds himself with only a cursory knowledge of an interesting subject that still remains outside his own work environment. Even when a publication can be found that relates TQM to the engineering environment and gives some idea of how to measure what has been accomplished, the reader is still left with no idea of how to go about implementing the effort.

To help fill this void, this book explores the concept of Total Engineering Quality Management. The manner in which TQM from the production/manufacturing environment can be modified, implemented, and measured within the engineering project environment will be presented and discussed in depth.

Ronald J. Cottman

CONTENTS

TOTAL
ENGINEERING
QUALITY
MANAGEMENT

INTRODUCTION

Product quality comes from the team working together. When implementing Statistical Process Control (SPC), the dedication of every person in the implementing organization is required if the quality of the process is to be monitored as well as measured.

SPC SHOWS TEAMWORK WORKS

> ### The Software Quality Assurance (SQA) View
>
> Careful Kevin, the tough software manager of the Pinafore Project, was addressing his troops. "You guys are the best S/W [software] engineers in the business," he said, his teeth clenched tightly. "I want no errors found when this code gets to the test weenies. Find and remove those bugs now, before we release the code to SQA or the S/W testers. Our goal is zero defects."

Let's Hear It for Zero Defects

"OK people, listen up and listen good." This was Fast Freda, the SQA manager, speaking to her section. "When you guys work with the Pinafore people, I want nothing but cooperation. Stop errors before they begin. Find them early and have them corrected up front. That's our job." She stopped speaking, placed her hands on her hips and looked directly at her group. "When we get this stuff for final SQA," she said with a steely glint in her eye, "I want no bugs in it." She turned on her heel and left the room. "Our job is to break this code under test," Timely Tim, the S/W test manager, said. He went on, "SQA is good, but not good enough to assure us error-free code." There was a cruel grin on his face. "Light off that VAX and let's get to it. Expose the defects. We're testing to show the presence of bugs, not their absence."

Boring Charts

Rabid Robert, the project person in charge of statistically charting the process of S/W production, was busily making small dots on a chart that hung on the wall of his office. Rabid Robert had been tracking the Pinafore Project S/W since the production of the first module. He was very disappointed. No defects extended above or below the control limits that had been established for the project code and his dots seemed to be grouped straight across the chart around the center line. He hated that. No one was making the errors he was watching for. The S/W test engineers were finding that the code was doing exactly what it was supposed to do. He reported his findings to the VP of Engineering.

The Winners

Wil Watt, the Engineering VP, had called the Pinafore S/W, SQA, and S/W test engineers together in his office. Pinafore had been completed on schedule and within budget. He leaned back in his plush chair and spoke expansively. "I have just received the report on your project from the SPC people," he said, placing his hands behind his head, "and I want you to know that I'm proud of our team." He smiled broadly and several of the people present nudged each other with an elbow and grinned. "You've joined together and turned out zero defect software. That's amazing." He paused for effect. The engineers moved forward on their chairs. "Therefore," Wil Watt continued, "I want each of you to take the next week off with pay."

Each person in an organization being tracked by SPC methods must recognize personal involvement. SPC can show when a process is out of control; however, SPC cannot ensure that a process remains in control. Process control is brought about by individuals.

Industrial quality assurance concepts based on past principles and policies are being closely scrutinized today by companies of every size. Changes are being instituted largely due to a cultural change in corporate thinking. Now there is recognition of the principle that quality cannot be achieved in a product without quality being achieved by each individual, in each organization, within each division of the company. An innovative approach to Quality Assurance is now being advocated in industry and in government. It stresses the education of all personnel in the policies and practices that encompass the measurement of the organization's quality effectiveness. Total Quality Management (TQM) programs are now being based on the ongoing update of quality processes to ultimately bring about the continuous improvement of the process.

Total Engineering Quality Management can be introduced into a strictly engineering operation, and all of its benefits can be realized. The same basic TQM rules apply to the engineering environment as apply to any other environment, and the same support group is required to ensure success. Management of a project is a process, and the success of any process equates directly to the manner in which the process is monitored and measured. Therefore, it is safe to say that before quality can be introduced into a process, good managerial techniques must be instituted and followed. Once those techniques are in place, the process can be instituted, and measurement can begin. Once the measurement devices are in place and metrics have been obtained, process improvement can be achieved. Each of these disciplines is discussed in the following pages. The discussion moves from Total Management of Engineering Quality to Statistical Process Control, and on to a method for measuring process capability. This text will provide the strategies for implementing a TQM program for the care and feeding of the engineering process. The statistical discussion is kept to a minimum and includes only the statistics involved with those process control items included in the text. However, if a more in-depth discussion is desired, I highly recommend a text called *SPC Simplified: Practical Steps to Quality.*[1]

It is with pleasure that I acknowledge the dedicated assistance of my dear friend Jack Alanen, Ph.D. Jack is a senior member of the technical staff at Litton Aeroproducts Division, and an extension professor of engineering and computer related sciences at the University of California, Los Angeles. From the start of this manuscript, Jack encouraged me and remained available for discussions and questions. Jack is a tough editor. I also acknowledge the assistance of Tim Hansen, the fine engineering administrator at

APD for his help with management chapters; my secretary and friend Leslie James, who checked the readability; and my wife, Cele, who assisted me in more ways than I have space to mention. Thank you. Thank you for your support and patience.

REFERENCE

1. Amsden, R. T., Butler, H. E., and Amsden, D. M. *SPC Simplified: Practical Steps to Quality*. White Plains, NY: Kraus International Publications, 1986.

1

MANAGING TOWARD IMPROVEMENT

Total Quality Management (TQM) as practiced in the United States is still in its infancy. Management of engineers and engineering organizations, on the other hand, is a practiced art with a solid foundation built over many years. Total Engineering Quality Management (TEQM) is a combination of the two. The tendency, therefore, is to pursue establishing a TQM plan just to be among the front-runners in a new field, and to ignore the key idea that before the Q in TQM can be brought about, the M must be brought solidly into the game. In other words, the basic foundation of management principles must be studied and implemented in any engineering environment before success in TEQM can be logically expected. It is virtually impossible to place too much emphasis on the study of the persons whom the engineering manager is going to manage. Without an understanding of the engineer and the engineering personality, the Total Quality program is doomed to failure.

An engineer needs to feel free to pursue his own ideas to a point of completion. Problems can result if an engineer is made to feel he is being closely supervised. Rules and regulations other than the laws of physics are a burden to the engineer that he would just as soon not be bothered with. Although he recognizes the existence of regulations, there is a tendency among engineers to bend those rules as far as possible without experiencing an actual breach.

To effectively manage engineers and engineering processes, a manager must place himself in the same logical surroundings as an engineer. In some cases the engineering manager has functioned in an engineering atmosphere for years. He is aware that an engineer's training requires him to be unwilling to accept rules unless some logical reason can be seen for doing so. It stands to reason, then, that to effectively control an engineering process, the manager must understand the engineers who are a vital part of that process.

THE ENGINEER TODAY

The engineers of today look somewhat like the following: they resent administration; they have a strong desire for freedom in work; they are preoccupied with detail, technically oriented, well educated, and perfectionists; and they are logical, take pride in their work, and have a great respect for competence. They expect that people will be as predictable as physical laws. How then do we manage processes that exist in the same environment as this complex personality?

STATISTICAL PROCESS CONTROL FOR THE ENGINEER

The engineering atmosphere is well suited as an area in which to apply process control. Statistical Process Control

(SPC) is accepted by engineers as part of the mathematical discipline to which they are accustomed.

Although a difference may exist in responsibilities and specific goals, every engineering process manager will need to perform the identical basic functions if he is to be effective. The process or processes that will lead to completion of a project on schedule and within budget must be selected. The long- and short-range goals of these processes must be carefully explained in detail to the engineering personnel.

The engineering manager who wishes to motivate engineers to use SPC must make that use a part of the engineering process. In a case in which dealing with SPC becomes an additional task, the engineer will tend to resent it.

Engineering Manager has SPC Success

Minnie the Manager and the engineers whom she supervised had been working together for more than five years. She was respected by each of them as an engineer who had come up through the ranks and who knew her stuff. Last week, Minnie was instructed by the VP of Engineering to implement a program of SPC into the lab test flow. The failures of a ramp generator card being integrated into a new system were to be charted, tracked, and traced, and the source of the errors corrected. Minnie knew she had a problem. To insist that her guys stop in the middle of a test to chart the faults of a failed device was more than she could bring herself to do. Not only that, but she didn't believe that she would have done it when she was working in the lab. What to do? Minnie left the VP's office shaking her head . . . and then it came to her. Project Engineering was already using SPC to track budget anomalies. The test engineers were already logging card faults into their engineering notebooks. Minnie the Manager had only to convince Useful Ulysses, the cognizant project engineer, that Racing Rhonda, the project analyst who worked for Useful and was tending to project SPC budget charts, could spend a few more moments at the end of each day charting the faults listed in engineering notebooks.

Minnie walked down to the water fountain where she knew Useful Ulysses the project engineer spent much of his time. She smiled gently and presented her problem. Useful Ulysses was, of course, delighted to offer assistance, and Minnie was "home free." Racing Rhonda was overwhelmed with joy at having an opportunity to be of assistance by adding this small task to her work schedule, thus proving beyond all doubt that . . . TEAMWORK WORKS.

In the above scenario, Minnie sets a good example of handling SPC implementation. When process control through the use of statistical charts must be initiated, not only must these processes remain flexible, but the manner of implementation must be compatible with those involved.

COMPARISON WITH OTHER COMMUNITIES

Engineering project managers, in establishing controls to bring the various processes of their project to a successful conclusion, must also be aware of the processes in related departments with which they must interface. A comparison of the engineering manager and a line manager in some other environment points out fundamental differences. The line manager's task consists mainly of directing the activity of those employees whom they manager. They expend a great deal of time evaluating personnel and the circumstances surrounding the use of personnel. In comparison, the engineering manager should pursue his managerial responsibilities in a somewhat different manner. The highly individualized engineers within the engineering manager's section may resent direct advice from the manager. Were evaluation of an engineering process required, then the engineer would expect to personally evaluate the process. While the manager may well be a fine administrator, failure can be expected if the sensitivity of the engineering personality is ignored.

PSYCHOLOGICAL CONSIDERATIONS

Understanding the emotional complexity of the technical mind can be of great assistance to the engineering manager. Having worked in the engineering environment for years, the manager already understands those who are working for him. However, an understanding of general managerial psychology can be helpful as well. The following insight was brought about when Douglas McGregor, in his excellent book *The Human Side of Enterprise*,[1] examined the assumptions normally made about people at work in various organizational systems. He felt that the basic assumptions of owners and managers greatly affected the way they went about their management tasks as well as the way organizational structures were built and run. On one side he placed the following assumptions associated with the personality he classed as the Theory X person:

- The average individual dislikes work and tends to avoid it if he can.
- Most individuals have to be pushed, directed, controlled, and threatened with dire consequences in order to get them to do enough to carry on with the work of the organization.

On the other side, McGregor labeled these assumptions as associated with the Theory Y person:

- People do physical or mental work as easily as they play or rest.
- People will show self-discipline and direction when

they are committed to the objectives of an activity.
- Commitment to objectives is related to the rewards associated with their achievement.
- The average person in a proper setting not only learns to accept responsibility but also seeks it out.
- The capacity to contribute imaginatively and creatively to solving the problems of an organization is widely distributed among people.

Understanding the McGregor study will enhance the engineering manager's overall capacity to deal with even the most complex personality and thereby provide a more satisfactory scenario for employee interaction.

Engineering Manager Splits Adams

Adamant Adam was a fine example of McGregor's Theory Y. Not only was work as natural to him as play, he easily accepted responsibility and displayed creativity. Adamant Adam was a dandy worker who did one-and-a half days of work every day. Adamant worked in the same software lab as a co-worker who shared a portion of his given name. His co-worker was a big, strapping guy named Indolent Adam.

Indolent Adam was a shining example of the Theory X personality. Not only did he avoid work, but it was said of him that he was "born lazy." Responsibility was to be avoided like the plague and he had to be threatened in order to restrict his coffee breaks to less than an hour. He performed only enough work to keep his job. However, Indolent was one sharp programmer when he felt like working.

Mickey Einstein, the new software manager, was becoming painfully aware that the work produced by this team was being written by one person. There were just no differences in style. He called Adamant Adam in and asked pointed questions about the manner in which work was being accomplished, and Adamant gave all the right answers. Mickey called Indolent Adam in and asked the same questions. There were very few answers forthcoming. Mickey knew he had a problem. He had to motivate Indolent Adam. Mickey started his campaign in a time-tested manner. He called in the employee who worked closest with Indolent to get some feedback on anything Adamant Adam may have noticed in conversation with Indolent which might indicate apparent problems. And Adamant unloaded.

"Darn right," Adamant said, "he feels that he is underpaid and you aren't interested in what he does." Mickey was shocked. Were his managerial skills slipping? "Not only that," said Adamant with feeling, "all of us in the lab feel like we're in a closed box. The lab is badly lit, rarely cleaned up, and the chairs are uncomfortable." Mickey stood, thanked Adamant for his feedback, and Adamant went back to work. "OK," said Mickey to no one in particular, "I can fix that stuff. Certainly Indolent is worth the consideration. But I don't think those things encompass the whole problem." Then the light went on in his mind. Use the motivational factors he had studied in school.

Mickey the software manager began going into the lab each afternoon to go over the work produced by his team of two. Indolent was somewhat uncomfortable at first, but as Mickey praised the team for their efforts it was just as apparent that the praise was enjoyed equally by both members of his team. He went one step further into his good management processes and began to assign specific, trackable tasks to Indolent. Further, he increased the employees' authority to make changes on their own and was pleased to see that the added responsibility was accepted, and in fact even relished. Mickey had not only solved his problem of an indolent employee, but had gained managerial knowledge through experience.

FEEDBACK

Interaction between an engineering manager and his employees can take on one of several names (teaching, conversation, counseling, training, networking), but every interaction that deals directly with the job scope of an employee or with the structure of a problem can be classified as *feedback*.

Feedback is the method by which one person influences another to change or modify an idea or behavior. It is the communication between two persons (or even groups) that indicates to each of them how they are affecting the other. Feedback is necessary if a team leader is to maintain the proper direction of the team's efforts in achieving a goal. In whatever form they receive it, feedback enables astute leaders to become cognizant of those things that either satisfy or dissatisfy the team.

MOTIVATION

In a study of the attitudes of people as they viewed their jobs in general, Frederick Herzberg and his associates came across these interesting facts and reported on them in a fine book entitled *The Motivation to Work*.[2] The information they gathered points to two major sets of factors that influence the motivation of people at work: hygienic and motivational.

Hygienic Factors

The Herzberg team concluded that few people are motivated by what they called hygienic factors, but most people would feel dissatisfied were they not present. Hygienic factors include the following:

- Receiving a salary increase
- Having good relations with supervisors
- Having beneficial personnel policies
- Having a prestigious job
- Having a competent supervisor
- Having good relationships with subordinates
- Having a secure job
- Meeting family financial needs and expenditures
- Liking one's co-workers
- Having good physical and social surroundings

A look at all these factors clearly shows that they are related to the environment in which a job is performed.

Motivational Factors

The items that compose the second set of factors, which they called motivators, are said to build a high level of job satisfaction. The team noted, however, that in most cases the absence of motivators did not create a feeling of great dissatisfaction. Motivators include the following:

- Seeing the results of one's work
- Receiving praise and recognition
- Performing creative work
- Receiving more responsibility
- Receiving advancement
- Having the possibility of growth in skills

The Engineer's Expectations

Motivators are related to the performance of the work itself. There is an obvious connection between these and a study performed by Abraham Maslow[3] on self-esteem and self-fulfillment. The Maslow study explains why the well-educated, complex engineering personality is far more motivated by recognition for his work than by salary increases. Money is important, but recognition is imperative and must be considered an important part of each engineering manager's approach to his personnel.

MASLOW'S HIERARCHY OF NEEDS

Maslow's Hierarchy of Needs (Figure 1.1) provides the process manager with a set of guidelines that, when applied, enable insight into an employee's present expectations.

Maslow states that the first consideration a person has on the road to satisfaction is of his physiological needs. When that person is uncomfortable physically, from hunger, temperature extremes, or even illness, his reaction to the satisfiers his environment provides is negative. Once the physiological requirement has reached satisfactory levels, this subject is ready to proceed to the next plateau in the hierarchy. Now he strives for safety.

Safety Factor

The importance of safety varies from person to person, but not necessarily from group to group. It was found that no more importance was placed on physical security by women

Figure 1.1   Maslow's hierarchy of needs.

Self-actualization
Ego needs
Social needs
Safety needs
Physiological needs

than by men. The degree of importance placed by an individual varied greatly, regardless of sex. Immediate safety (physical) and immediate security (financial) were necessary components in the hierarchy before the subject was ready to proceed to the next rung on the ladder—love.

Social Factor

An employee needs to feel that he belongs, and the alert engineering manager will provide an environment compatible with that social need.

Maslow places love (affection) in the third position of needs, and weights it heavily as a satisfier. Negative feedback from the process manager to the engineers can be a stumbling block in the area of affection and must be handled with great delicacy by the manager. The social needs of the engineer must be kept uppermost in creating an atmosphere of "belonging," and the manager must be aware that feedback can be a destructive force when it serves only his own needs and fails to take into consideration the needs of the other person. Negative feedback must not be directed toward behavior or circumstances over which the person addressed has no control.

Ego Factor

Once the needs of these physiology, safety, and social factors are met, a person will seek to satisfy his need for self-esteem in direct proportion to his ability to accept his present level of achievement. This may be an almost impossible level for him to reach. In such a case engineering managers must once again rely on feedback to alert them to a potential employee problem. The feedback can come from the person with whom the manager is dealing and also other persons in the engineer's immediate vicinity. By using available feedback and reacting to what is needed to assist the employee in realizing ego reinforcement, the manager assists the employee in attaining this important level. The feeling of self-esteem provides the ego reinforcement that enables the employee to strive for even better positions in the advancement spectrum.

Feedback must be solicited, and is most useful when it is received directly from the person involved; however, clear, concise information from the general area of concern is of great assistance no matter what the source. A manager may be assured that he has succeeded in the development of a positive ego advance in an employee when it becomes apparent that the person likes himself and demonstrates this feeling in his general attitude. Once this development has been obtained, that person will then reach for the highest step in the Maslow Hierarchy.

Self-Actualization Factor

Self-actualization can only be found through self-fulfillment. Maslow has placed this last in the needs of a human being because until the other needs have been met, self-fulfillment can receive no focus. The manager must realize that this particular need is a dynamic one. The need is characterized by both verbal and nonverbal feedback and must be both biological and physiological in order to be complete. Self-fulfillment is not the last step in an engineer's development. It is merely the last rung on the ladder of his hierarchy of needs. Once he finds himself fulfilled in his present position, he will then accept that position as the bottom of another ladder and move again through the hierarchy toward an even higher position.

SUMMARY

Total Engineering Quality Management requires that each process within a system be under constant examination. As Maslow points out, the process of an employee performing his daily tasks is actually a system of emotional considerations that can lead to either success or failure. By closely observing (even tracking) the individual through the various steps of the Maslow Hierarchy, an engineering manager can anticipate personnel problems and open the lines of communication that will preclude problems from happening. This type of action by the manager is the basis of TEQM. Make the correct changes to the process and problems begin to diminish.

2

PROCESSES AND SYSTEMS

A SYSTEMS APPROACH TO ENGINEERING PROCESS CONTROL

The concepts and theories dealing with processes and their control have seen expansive changes in recent years. With the advent of computer control of data systems for management information, the integration of external data into the engineering information system has been optimized. Now, through the use of Management Information Systems (MIS), data from diverse fields of knowledge can be combined to ascertain the effectiveness of any particular activity. In the field of quality control, modern developments have made it possible to maintain control over quality by applying a management approach that looks at the total picture rather than only analyzing the individual components. Since process control is an outgrowth of quality management it is only appropriate that statistics be applied to the search for quality in engineering processes and applications.

MATHEMATICS: A COMMON LANGUAGE IN SYSTEMS THEORY

In a test concerned with general systems theory, Kenneth Boulding[4] in 1966 addressed some of the problems that stood in the way of a systems approach to continuous process improvement. Dr. Boulding noted that within a process structure each functional team had its own language. He therefore advocated the common language of mathematics as a standard notation that could be successfully utilized. Not only could mathematics be used for integration, it could also be used as a measuring device to achieve understandable data. Statistical Process Control (SPC) was born without being named.

Systems Improvement

The concept of continuous process improvement had begun. Consideration of process improvement as applied to the control of engineering quality became a requirement. The world of quality assurance found that by measuring the dynamics of the observed variation in an organization as well as in a process, the process could be continually monitored and improved.

Organizational Dynamics

Organizational dynamics were presented by Johnson, Kast, and Rosenweig about ten years after the Boulding study in their popular book *Theory and Management of Systems*.[5]

Their studies compared the dynamics of a corporate organization to an open-ended cell.

They began with the recognized premise that a cell is an open system that maintains a constant state while matter and energy that enter it keep changing within. The cell is influenced by these changes, and it influences the surrounding environment accordingly. Ultimately, a state of dynamic equilibrium is reached by the cell within the constantly changing environment.

This description of a cellular system adequately describes the condition and environment of the typical business organization. The business organization is also subjected to dynamic interplay with its environment. The environment of a business organization is made up of customers, competitors, labor organizations, suppliers, government, and other agencies. To carry the comparison further, the business organization is also a system of interrelated parts working in conjunction with each other in order to accomplish a number of goals.

Development of the Model

Following this comparison, the three scientists went on to develop a model whereby the nervous system of the body was held as a communication system, and the brain served in the capacity of upper-level management. Their model was popular, formed the groundwork for some of today's management concepts, and is still valid.

To bring about improvements within a constantly changing system requires tracking and measurement of the changes. Statistics have become the major tool utilized for this purpose. Statistical control methods applied to the various aspects of a business organization can provide the data that will demonstrate the efficiency of the organization as well as the individual processes within the organization. Once the method of sampling is identified, SPC charts can be formed to identify those particular points in the organization that are out of control. The process being examined can then have the out-of-control points worked individually and brought into control.

The Keys to Efficiency

The keys to an efficient system of process control are communication, commitment, visibility, speed, and simplicity. The more unsophisticated the system can remain, the more easily the steps may be monitored.

WORKPLACE APPLICATIONS

There are usually two processes within the workplace. One is the process that is stated in company policies and procedures, and that management believes is in place. Then there is the process actually being practiced. The two do not equate. A process developed by management personnel is conceived from past experience. A process that is developed by engineers who are doing the current work within the laboratory is a state-of-the-art process.

An excellent example of this is seen when an engineering manager is hired from an outside source and brings with her the laboratory policies of the past employer. The procedures with which she has been dealing in her last job are good ones and she can see how their implementation will improve the laboratory in which she now finds herself in charge. The problem however, is that the procedures she advocates are diametrically opposed to the laboratory practices at her present company. Could this happen?

Good Gilda Goes for the Gold

Good Gilda walked into the lab this morning and could feel the tension. Top Tom, the total test pterodactyl, was scowling at a green monster on his work surface. Jumping Jack was standing by nervously holding the cord that appeared to tether the monster to an ominous black box on a shelf just above the work surface. Thoughtful Thalia was standing with shoulders hunched forward and hands thrust deeply into her lab coat pockets. She too scowled at the green monster. Good Gilda had been in charge for one week and had an immediate feeling that she was about to be tested.

She walked boldly up to the group, smiled, and said, "How goes it, Top?" There was silence. "Looks like you have yourself a connector problem, eh?" Did she hear a low growl, she wondered? "Well," she sighed, "I guess you guys have got it all in hand." She didn't want to appear pushy, so she turned to leave.

"Wait a minute, Good," said Top Tom, "We really need to talk about this." He took the cable from Jumping Jack and waggled it in front of Gilda's face. "Intermittents!" he said with feeling. "These darned things have intermittents. More intermittents than you can shake a stick at. And I'm sick of it."

Gilda had heard this story before at her last company. "What replacement pins are you guys using on those things, Thoughtful Thalia?"

Thoughtful looked up slowly from her meditations on green monsters with intermittents and replied in a slow and deliberate tone, "Original equipment, Gilda. We replace them with original equipment. Good old steel. Won't bend, won't break." A weak "won't conduct," came from Jumping.

Gilda knew exactly what should be done. Gold pins had to be used for pin replacements to ensure a full interface at the female side of the link. "I'll tell you what, guys. I know policy here says use original equipment, but we've got a turnaround to meet. Thalia, have a Material runner get over to "The Pin Place" on Rustic Boulevard and buy 400 gold-coated replacements on an open-purchase rec."

"Unh-unh," the trio chorused, "That's not policy here."

"Then we'll just have to try and have it changed," she said. "Right now, we've got a turnaround to meet and we need state-of-the-art stuff in this lab. I'll take the responsibility. Thalia, you get on your horse."

Thalia lifted her shoulders and smiled. "If it works, you're a genius." She went off to Material.

Engineering Participation

With the advent of Continuous Process Improvement (CPI) techniques, participation by employees in process development has become important. The CPI method allows individuals as well as teams to improve the processes they deal with daily. There is a consideration of employee experience that enables an evolution of excellence rather than a revolution of engineering. One of the most important ideas to come out of the development of process control focuses on quality being completely process oriented rather than goal oriented. Total Engineering Quality Control enables quality to be engineered in at the beginning of a process, rather than encouraging an attempt to inspect quality in after completion of a build.

System Approach Aspects

A system approach to process control will ensure that improvement is a part of "doing business as usual." By including all engineers in process development, a team attitude is brought about and a change begins to take place. Recognize that the change will be a slow one; much time is consumed in bringing engineers together for conversation and idea exchange. Much time is spent in problem identification. Before a failure in a process can be repaired, the reason for that failure must be identified.

Changes Start with Management

Change begins slowly. There is a natural reaction in the workplace to treat the symptom (cause) of a problem rather than to go to the resultant effect and improve the process. The tools for process repair (process control) are new to today's workplace and need to be taught. Engineers who have reacted as "firemen" to problems for years will have a difficult time in changing their philosophy from reaction to planning. Process control charts and the various control devices will be discussed in later chapters of this text; however, recognition must be up front—that there is no quick method to bring about a change in philosophy.

Fourteen Points to Success

Dr. W. Edwards Deming is widely recognized as a teacher of quality management, and is known throughout the world for his "Fourteen Points for Management." His teaching of quality and process control dates back to Japan after World War II, where he was commissioned in 1951 to lecture Japanese manufacturing companies on the virtues of Total Quality Management (TQM). Japan awards a "Deming Prize" each year to the Japanese company that most widely demonstrates Deming's principles. Dr. Deming's teachings are condensed into fourteen major points and presented along with a short biography by Ronald Christofono[6] in his workshop series. The series is a fine discussion of process improvement and is recommended reading for those who wish to carry their studies on the subject further.

Dr. Deming teaches that in order to succeed, all fourteen points must be regularly applied. Although they are not agreed upon by all people, the fourteen points are presented here as an excellent starting point for any quality program.

The engineering manager may find a problem in adapting all of the following points into his or her philosophy. When one of the points cannot be justified, adapt the idea to existing conditions and develop the concept.

1. *Create a constancy of purpose for improvement of product and service.* A commitment to quality improvement is required. Management must have a long-range vision that will ultimately be based on continuous improvement of processes.

2. *Adopt the new philosophy.* We are in a new economic age. No longer can we accept the delays, mistakes, and defective workmanship that now exist in the workplace. Worldwide competition has introduced new competitors and also different means of obtaining a competitive advantage. Customers now expect that the producer will provide the market with excellence.

3. *Cease dependence on mass inspection as the primary method for improving quality.* Require instead that statistical evidence be supplied that quality has been built in. Anything less is costly and will produce a higher price for the consumer to absorb. Look for the method to accurately obtain the evidence by examining the causes of variation within processes and striving for process improvement through teamwork and employee participation.

4. *End the practice of awarding business on the basis of price tag.* Price does not necessarily equal cost. An initial low price could easily turn into a higher consumer cost after all costs are considered.

5. *Constantly improve the process of production and service.* It is management's job to work continually on the system, as well as to find problems and develop opportunities for their solution. There are only two sources of problems: processes and people. Deming says that only 15 percent of quality problems rest with the employees. That leaves the rest to be the fault of the process.

6. *Integrate modern methods of on-the-job training.* Training must center on job location and correction of process variation; anything less is only a temporary fix. By centering on the correction of variation, it becomes logical to teach the tools of SPC.

7. *Develop tailored methods of supervision and management.* Strive for quality in supervisory personnel rather than great numbers of micromanagers causing delays in the workplace. Supervision must be by example and by demonstration, and must focus on the commitment of the supervisor to improve process control.

8. *Drive out fear.* Effective work cannot be accomplished under fear of derision or punishment. Communication must be encouraged to be a "two-way street." Feedback from engineer to engineering manager and from manager to engineer must be accomplished. The basis for continuous process improvement is cooperation and teamwork at all levels with objectives and incentives shared by both engineer and engineering manager.

9. *Break down barriers between departments.* Feedback, feedback, feedback. Communication, communication, communication. When importance is placed on the interfaces between various departments, communication is a natural outcome. Barriers are removed and interdepartmental cooperation occurs.

10. *Eliminate slogans, numerical goals, posters, and other pressure-creating devices.* Process improve-

ment through employee participation will occur when new levels of efficiency are asked for without the method being provided by management. Improvement must be encouraged through the individual engineer.

11. *Eliminate procedures that require a specific output from each employee.* Instead, concentrate on the formation of a team attitude within the laboratory. Procedures that require a specific numerical output by an individual engineer will ultimately produce shoddy workmanship and create an atmosphere of error.

12. *Remove the barriers that stand between the engineer and his right to pride in workmanship.* When an atmosphere of teamwork is presented and maintained, an employee will know just what is expected of him. Communication between work force and management is kept at the maximum and the engineer's satisfaction with his job is at the highest level.

13. *Institute a vigorous program of education and retraining.* Enable each employee to function within a team of peers. Provide an atmosphere through education that is conducive to maintenance of dignity and satisfaction in the laboratory environment.

14. *Encourage every individual within the workplace to dedicate himself to this transformation.* Reward those engineers who support the new system, and concentrate on the development of policies that are system oriented. Make haste slowly. Allow the required changes to take place over a long period of time. The best results occur when the new method merely replaces the existing one. The two systems run parallel for some period of time, until the old method just fades away.

A Deming Critique

After studying Deming's fourteen points, it is well to look briefly at his critique of management. He labeled the critique the "Five Deadly Diseases" of a process. Deming feels that when a project is subjected to the following five conditions or activities, it will begin to exhibit signs of sickness and ultimately succumb:

Deming's Five Deadly Diseases

1. Lack of constancy of purpose
2. Emphasis on short-term profits
3. Evaluation of performance; having a merit rating system, or an annual review
4. Mobility of management
5. Management by use of visible figures only

Opulent Ollie Opts to Operate Ugly

Opulent Ollie relieved Likable Len as Vice President of Engineering on Friday, July 12. It was now September 16 and things had gone steadily downhill during the past five or six weeks. Of course, Ollie already knew exactly why there were projects behind schedule and over budget; it was because the laboratory managers were babying their engineers. However, just to show the engineering directors that he was willing to listen, he'd ask their opinion at today's staff meeting. The meeting would begin in just five minutes and he walked in and sat at the head of the table in the Engineering conference room. He leaned back in the chair, folded his hands across his ample middle, stuck his legs out straight under the table, and waited for the directors.

Design Don walked in and sat down without saying anything. He was about half way down the table. "Hey, Ollie," System Sam grinned as he sat at the far end. Project Paul, who headed up Project Engineering, came in with a scowl (he was responsible for tracking budgets) and said, "OK, let's get started." Software Sally came in, nodded to each person, and sat back to sip her coffee.

"OK," began Opulent with no preamble, "we've got schedule slips and budget slips. You guys are the senior managers of this department. Tell me what's wrong!" He stopped speaking and glared around the table.

There was some shifting around in chairs, and then. . .

"OK, Opulent," began Software Sally. "It seems like canceling paid overtime was the wrong thing to do. The policy that was just put in about three months ago paid the guys for their efforts, and now we've changed it. They just don't like it. I think we have a rebellion on our hands." She picked up her cup and sipped.

"I'll tell you this, Ollie," said System Sam. "Engineers just don't like vacillation in the policies they work with. I think we ought to go slow on changing the things that Likable Len had established.

"The guys in my labs complain that we don't seem to have any purpose or direction anymore. They say there are too many changes."

Opulent's face turned the color of a stoplight. His voice was raspy when he spoke. "You listen to me," he began, his voice a growl. "No constancy of purpose?" Ollie was visibly attempting to control his anger. "I'll give you constancy of purpose." He glared at each person in the room. "Maximize profits," he shouted. "That's constancy of purpose. Work overtime for no pay when there's something to do. That's constancy of purpose." He was shouting now. "We maximize profits forever. That's constancy of purpose." With that, his mouth snapped shut and he leaned back, breathing deeply.

"But Ollie, engineers don't even think about profits. They think about science and technology." This from Don Design. Don was noted for his egghead attitude.

"I'll tell you what'll make 'em think about profits. Cut out a holiday next year. Kill 'em at merit review. No raises. That'll make 'em think about profits." He was waving his arms while he spoke. Now he leaned back in his chair again.

Paul leaned over and whispered to Sally, "The floggings will go on till morale improves." She turned slightly and nodded in agreement.

Ollie began again. This time his voice was somewhat controlled. "You guys get out of your offices and into those labs. We need you in the trenches. We need you doing what you're paid to do. We need you managing. Change out slow workers. Get rid of the gripers. Replace your managers. They're really the ones at fault here anyway. Change the things that I don't like!"

Project Pete finally found his voice and said quietly, "Opulent, those guys just aren't going to like it. That's not what they're used to. They want to be left alone to do their job, and paid for doing it. We can't manage-in fidelity to the company. We have to provide real incentive, if things are to get back to where they were."

Ollie concluded: "You listen to me. The president tells me what to do and I tell you. Now go do it."

3

THE PROCESS AS A SYSTEM

TYPES OF PROCESSES

A process is defined in *Webster's Student Dictionary*[7] as a series of actions to a given end. A process then is a series of steps, which when placed in motion, will accomplish a specific goal.

An example of a familiar process is that of walking across a room. By taking several steps in the same direction, the goal of crossing the room is accomplished. All goes well with the process as long as nothing gets in the way of a step. However, when the steps of the process are inhibited, then the normal action must be changed. Different types of steps are required. At each inhibition, that particular step becomes an individual problem. In the case in which the person performing the process of crossing the room cannot decide how to get around that which is inhibiting him, outside assistance may be required. Persons who will lend assistance are brought together and begin to act as a team that will figure out the manner in which the journey across the room can be completed. The process can no longer be performed by a single entity, and with the addition of persons to the team, it becomes a system.

Returning to *Webster's Student Dictionary*, the definition of a system is found to be: "An orderly assemblage of facts, parts, etc." In the case of this process, the system is composed of people. The system is employed as a major tool of a project, and the process is just one leg of the system.

PROCESS/SYSTEM MANAGEMENT

Total Engineering Quality Management (TEQM), when applied to a complex process, may be defined as "attainment of a goal through the utilization of various disciplines working together as a team." Utilization of the talents of those persons who are not part of the engineering manager's immediate department requires that the manager establish an approach to process control based on procedures that are easily understood and highly visible to all departments involved.

Establishing Objectives

To attain the goal of the team, the procedures developed by the engineering manager will be the basis by which objectives are clearly defined and set. Each objective should be written and laid out in any format convenient and understandable to the process team. The objective need not contain the plan for success; however, the manager must establish early on the plan for the process team to follow. The team must see the plan in writing, and the manager must ascertain that each of the team members understands the plan.

Writing the Objective

Three steps can be followed to provide understandable written objectives as well as the activities that will ensure these objectives are met. Two types of objectives should be considered: developmental and evaluation.

The Developmental Objective

The developmental objective, and also the implementation activities and the establishment of evidence that the objective has been met, can be laid out as follows: (Note: These are examples only.)

Developmental objective #1: By July 14 all test error data dealing with the Franca Project will be collected and an evaluation form will be constructed through which each team member will have provided evaluation input.

Implementation activities: Data will be collected by a team composed of system and test engineers, and developed into check sheet form by July 9. The evaluation form will be prepared by Jones and Smith prior to July 10 and submitted to each team member for personal processing.

Evidence the objective has been met: The objective is met when the data have been evaluated by each team member and that evaluation form has been turned in to the process team leader.

The Evaluation Objective

Evaluation objective #1: By August 1 the process team members will analyze the data provided them and develop selected charts and graphs portraying the process in question. The charts and graphs will be used by the team to analyze the condition of the Franca Project hardware.

Implementation activities: The information gained from the charts and check sheets from each team member will be collected and combined by the team leader. These data will be used to lay out a run chart or an $\overline{X}$-R combination during the team meeting of August 5. The chart will be analyzed by the team members outside the meeting, and they will begin deciding on solutions to any problems they find.

Evidence the objective has been met: At the team meeting of August 12 the team leader will present to the team the solutions brought to the meeting by team members. The suggested solutions will be brainstormed for selection of a problem solution.

Objective Summary

The above scenario is a simple example of how objectives can be successfully laid out and examined. The example is by no means the only manner in which objectives can be written or used. It is suggested that the resources available to a team always be considered prior to writing objectives. In this manner, resources can become a part of the total consideration.

ORGANIZATION OF RESOURCES

Organization of resources is a step of the managerial process that is sometimes left until the project has begun. When this is the case, schedules are usually impacted by the lack of material, capital equipment, test equipment, or other critical items. Resources must be organized before the beginning of a process, and especially before the staff is selected. In this manner, staff members can be selected for their ability to work with available resources. Adverse impacts are brought about, for example, when there are no software development stations for the assigned programmers, when there are too few assets to be utilized by the test engineers, when Software Quality Assurance has insufficient staff to maintain schedules, when there is inadequate communication to the Human Resources Department of what is required, and especially when the scheduling of work handoff between teams is inaccurate.

The steps toward establishing Statistical Process Control (SPC) can be taken at this point and the controls needed for completion of the process can be brought about early on. Procedures and instructions can be issued and motivating forces for the staff can be put in place.

The process manager can apply those measures required to bring about any actions that will ensure a smooth process start. By remaining flexible at this point, the engineering manager can virtually ensure success.

THE PHASES OF AN ENGINEERING PROCESS

Much research has been conducted in the manufacturing and production communities on establishing the various phases a process goes through from beginning to end. An additional phase needs to be added when the search for total quality is applied to an engineering process. Quality Department personnel can be integrated into the process at the start, thereby relieving the burden of extensive inspection in the final stages of the process. It is generally agreed that the following four phases, which were conceived by D. I. Cleland and W. R. King in their book *Management: A Systems Approach*,[8] are those through which a process will progress.

- Conceptual
- Production
- Operational
- Divestment

However, for an engineering process, an additional phase must be added to complete the phase example. The added phase is called

- Lessons learned discussion

Adding a phase after completion that considers the lessons learned either by a successful or an unsuccessful process provides history that can preclude future errors and also provide positive examples of process management. Each phase will be considered individually in the following discussion.

Conceptual Phase of a Process

During the conceptual phase of a process, evaluation is performed to ascertain the validity of the ideas that the engineering manager plans to place in effect. The following procedures provide an orderly manner of accomplishment for this phase. The manager should

Establish process concepts that provide initial strategic guidance to overcome potential deficiencies.
Determine initial technical, environmental, and economic feasibility and practicability of the process.
Examine alternative ways of accomplishing the objectives.
Provide initial answers to the questions: What will the process cost? When will the process begin? What will the process do? How will the process be integrated into existing project efforts?
Identify the human and nonhuman resources required to support the process.
Select an initial process design that will satisfy all process objectives.
Determine initial process interfaces.
Establish a process organization.

A preliminary risk analysis is performed during the conceptual phase. Time, costs, and performance requirements are clearly documented and the decision is made as to whether or not the process can be pursued. In the case of competitive programs, the decision to bid or not to bid the contract is made.

Once the process time and performance requirements have been firmly established, a decision about cost can be made more easily. At this point, most costs of pursuing the process will be recurring costs. Recurring costs are, quite simply, operating costs. Personnel performance of a high quality done in an efficient manner can reduce the recurring costs considerably. Recurring costs are often cited in support of a valid training program begun prior to the process start.

In the conceptual phase of a process, final performance requirements are laid out and documented. This phase is the questioning phase. For example, a question often asked too late in a project is: How will we know when we are finished (or if the product is acceptable to the customer)? In other words, disregarding schedule, budget, technical constraints, and other problems (assuming they will be overcome), how will we know that we have designed and built what we need?

Identification is made of those areas of the process in which high risk and uncertainty exist, and delineation of plans for further exploration of such areas is accomplished. Interface to external support systems is established, and identification of the documentation required to support the process through completion is recognized (e.g., policies, job descriptions, budget and funding papers, and memoranda).

Looking toward the lessons learned phase, the engineering manager should keep a strict log tracking the results of each step taken during this phase. Properly instituted and maintained, the log will provide historical data for discussions of lessons learned following process completion, as well as indicating when to introduce the production phase.

Production Phase of a Process

The production phase provides the opportunity to test and standardize the process. It is generally recognized within the engineering community that the documentation identified as necessary in the conceptual phase must be finalized in the production phase. Once again, the steps of accomplishment should be carefully tracked by the manager, and consist of the following:

Updating the detailed plans conceived and defined in the conceptual phase

Providing identification and management of those sources required to facilitate the process (e.g., inventories, supplies, labor, funds, and budgets)

Verifying process specifications

Preparing and disseminating final policy and procedural documents

Determining whether or not the process is adequate to do the things it was intended to do

Developing affiliated documentation (or manuals) describing how the process is to operate

Developing plans to support the process during the operational phase

The operational phase follows the production phase and relies heavily on all work, as outlined above, being completed. The historical data log kept by the manager will provide information about when the production phase is complete and the operational phase can be successfully implemented.

Operational Phase of a Process

The operational phase integrates the process into the existing organizational system. During this phase the process output is introduced into the established engineering system. Noncritical personnel are trained to carry out such postdelivery tasks as training the customer and performing the technical tasks that will maintain the system in the field.

Operational phase details include the following:

Monitoring the use of the product produced

Integrating the process's product or service into existing organizational systems

Evaluating the technical, social, and economic sufficiency of the product to meet operating conditions

Providing feedback to organizational planners concerned with developing new processes or systems

Evaluating the adequacy of supporting systems

The operational phase provides the engineering manager the opportunity to integrate departments external to engineering into the process concept. Quality inspectors are required, the marketing department begins to generate interest, and communication between departments becomes imperative. Completion of production is the milestone that provides the kickoff of evaluation. Cleland and King call this phase the divestment phase.

Divestment Phase of a Process

The divestment phase occurs when a process is removed from the engineering environment. This may mean that the process is no longer a requirement for any particular project, or possibly because of a short-term halt in some particular phase of the project's life span. Reallocation of resources is activated after the decision is made about where they should be assigned. Evaluation of the process then takes place and serves as input to the conceptual phase of subsequent processes. To provide an orderly manner of transition for the process from activity to inactivity, the following process phaseout will provide an excellent entrance into the lessons learned phase:

1. Develop plans transferring responsibility to supporting organizations
2. Divest (or transfer) resources to other processes
3. Develop lessons learned for inclusion in a data base to include
    a. Assessment of image to customer
    b. Major problems encountered
    c. Technological advances
    d. New or improved management techniques
    e. Recommendations
    f. Other major lessons learned as extracted directly from the engineering manager's logbook

Note that this technique of engineering process management lends itself well to processes of any scope or magnitude. Experience has shown that this process life cycle occurs in processes small or large.

The Lessons Learned Phase

The research accomplished by Cleland and King was built around the accomplishment of the basic (generic) project. With the implementation of TEQM comes the responsibility for examining all engineering notebooks kept during the course of the process, but especially the engineering manager's log.

Through use of the lessons learned discussion, the engineering manager can reap the advantages that the combined knowledge of his workers can provide, and integrate improvements into existing and future processes. The lessons learned phase gives the team an opportunity to look at what was right, and also those things that may have gone wrong. Much can be gained by holding a postmortem on a successful project finished on time and within budget. Juran[9] suggests the appointment of a project historian at the launching of the project. As the project progresses through the conceptual, operational, and production phases the historian has the responsibility for maintaining an operations log. In the phase down of the project, when lessons learned discussions are held, the log can act as an agenda for the meetings. An excellent, although drastic, example of lessons learned can be seen in the successful recovery efforts of the pilot recorder "black box" following an aircraft accident. The recording can point to each incident leading up to (and in some cases what occurred during) the event. Using the recording as a guide, aircraft industry officials can minimize the chance of future accidents.

Disasters can teach what not to do the next time, but successes can lay down a path that leads to even greater successes. The historical log beginning at the project's conception can be instrumental in ensuring that operational phase rework occurring on that project is not required in future efforts.

4

THE ENGINEERING PROCESS TEAM CONCEPT

PROCESS MANAGEMENT

Management of engineering personnel during the scope of a process is important; however, management of the process itself is the key to continuous process improvement. The Continuous Process Improvement (CPI) atmosphere that currently permeates the engineering atmosphere, advocates process control by the process team.

Senior management has been made aware of the team approach to problem solving, and in most cases has come to realize that it can bring about radical improvement in any organization. In an engineering organization, problem-solving techniques based on the process team concept provide an synergistic atmosphere that is being accepted more and more by industry as the most efficient manner in which to provide complex and diverse input toward problem solution. In many areas in which teams have become important in problem solving, senior management has agreed to act as the program steering committee.

Process Measurement

Costs are reduced by controlling unanticipated variation within a process. When variations are controlled at the beginning of a process and then maintained throughout, inspection becomes a thing of the past. The result is lowered costs for process completion. This has proven highly successful.

The tools of Total Engineering Quality Management (TEQM) provide a method for monitoring and measuring that variation. This method has been used in Japan since around 1952 and has had proven success in the use of statistical charting tools that make possible the charting and description of process behavior.

Process Statistics

The method for measuring and monitoring process variations is called Statistical Process Control (SPC). This book describes only that portion of SPC required to monitor engineering processes. The description is by no means complete, and if more depth is required, there are many excellent books that go deeply into the statistical system for process control. Two highly recommended texts are the *DataMyte Handbook*[10] and again *SPC Simplified.*[11]

For engineering process control, charting methods (to be discussed later) are available to provide the required information and feedback for an efficient system of reducing variations in an engineering process. Through the use of SPC charts, a process can be measured and decisions can be made as to whether or not the process is stable, and whether or not the process can perform the tasks required of it in the first place. Study of the process capability through the use of simple statistics will provide the information to prevent process deterioration, and to give a true picture of the real power of the process.

The Backbone of System Processes

The backbone of the system that provides information on process capability rests in two statistical charts. One chart, the $\bar{X}$ chart, deals mainly with averages. The other, the R chart, deals strictly with the ranges over which the process functions. Through the use of these charts and the interpretation of the information they provide, the engineering process team will provide a process that is stable, with no variation from external causes, and that is based on the reduction of any variation required by its specification.

THE ENGINEERING PROCESS TEAM

The engineering process team can consist of two or more persons in accordance with the size of the organization sponsoring the team and the team's mission. The team concept most easily succeeds when the company makes a decision to effect a cultural change. This change occurs when training in team methods is made available for each employee, and management provides incentives and recognition of the team effort. In other words, the training effort must be such that every member of the company work force has the opportunity and desire to participate.[9] To provide an atmosphere of team participation in problem solving is to offer a feeling of solution ownership to the team participants.

Engineering Director Sez: Liz Knows Her Biz

Our study takes place in the test engineering laboratory of the Engineering Department. The test device storage area is surrounded by low, open shelves used by the engineers. They place the devices to be repaired

---

or calibrated into protective containers and place the containers on the shelf with a suggested priority for repair.

The devices to be serviced are then taken from the shelves on the other side by service technicians working in the enclosure.

Loyal Liz, super service supervisor, has been fielding complaints from various fronts. The engineers are complaining that their priority markings are being ignored. The service techs are complaining that the tags containing the priority markings fall off the protective boxes that are provided for the test devices the engineers stack on the shelves.

The engineers complain that the material of the boxes is such that the tags cannot be securely taped to them. The service techs complain that the shelves get stacked so full that even when the tags remain intact, they are covered by each other and can't be read.

Liz remained calm throughout the scene and told all those involved that she would give it thought and get back to them. She walked into her office, closed the door, and applied her grand intellect to the problem. In no time at all the solution struck. She would talk with the service director about sponsoring a three-member team consisting of Manufacturing Inspection, Production Control, and Test Engineering.

Great idea! The director agreed immediately. Here was the opportunity to get the problem solved, place the solution into effect, and have members of the three organizations involved come away with a distinct feeling of solution ownership.

Liz took the first step. She contacted the managers of the two sections, and since they were already aware of the problem, they were happy to assign a knowledgeable person to the team. The next step required a chat with the director of TQM to have a facilitator assigned to the team. DONE! Her team now consisted of three good employees and one facilitator. The team elected a team leader and agreed to meet each Thursday at two o'clock for one hour. Since all members of the team had taken the company TQM training program, they functioned with efficiency. They solved the problem.

The team made an interesting presentation of the problem and solution to the senior staff, and they received congratulations, budget, and permission to place the solution into effect. SUCCESS! An interdepartmental problem was recognized.

The team system had worked. A team had been formed. A solution to the problem was formulated by the team. A management summary of both the problem and the solution as presented to the senior staff. A budget was assigned. The solution was implemented.

LIZ KNOWS HER BIZ.

A Process Team Overview

The general course of doing business in today's highly technical and industrial society requires a sophistication in decision making that can only be obtained through communication. The application of many minds to specific problems can bring about solutions far more accurate than individual effort can produce. The engineering process team approach to those solutions therefore becomes a viable tool in the decision-making process. A team of interested engineers targeting a specific problem brings an orderly approach to its solution not ordinarily found in the general conduct of business. The problem is placed at the working level, in the hands of engineers who are in daily contact with the reality of their world.

An often overlooked tangible benefit of such an approach is the view of the working environment provided to upper-level management. In the team concept of TQM, a presentation to senior management is required of the team upon completion of the team's effort. The thoughts of a group of engineers rather than the opinion of a single manager is communicated. Senior management is made aware and kept aware of the problems encountered at the factory and laboratory levels.

Engineers, who are formed into a group of decision makers, can freely express themselves at the peer level. They are encouraged to be as freewheeling as possible. Ideas and concepts usually kept under wraps from the engineering manager are brought out for inspection by other team members. The engineer who would be reticent to open communications with a supervisor can act in the forum of an engineering process team without fear of ridicule.

A + B + C + D + E Does Not Necessarily = F

The scene opens as Al calls Bob over to the side of the engineering lab

Al: Bob, would you ask Chuck to call Don and ask him if he'll have Evelyn tell Elwood to fix Fred's problem. I don't know why, but Fred complained to my boss that he has trouble communicating with Elwood.

[*Curtain*]

The group of employees formed into a team can begin to function as an entity and bring to light those things found in the course of their daily tasks to be helpful. The weekly meetings of the team become a place to release energy and gain recognition as a group, not just as individuals.

TEAM PARTICIPATION IN PROCESS CONTROL

A process can be defined as a set of given actions culminating in some conclusion. A successful conclusion usually results from a controlled process. The input into a process team by a motivated engineer will provide valuable information to control the process under deliberation. Technical knowledge from the level of individual members is brought to bear on the process and the parts of a process are separated into those areas coincidental with the knowledge of the team members. Open participation provides a spectrum of knowledge encompassing the total learning of those employees selected for team assignment. Benefits to the organization are gained through significant contributions by employees who would not otherwise be heard from.

Organizational Benefits

The process team provides a win-win atmosphere. The organization benefits from the efforts of the team members to an extent not realized from individuals. The employee benefits by being in a position to directly affect and influence his own environment. Those things that are usually accepted as being beyond the control of the working engineer are brought within reach, (e.g., work area, management communication, and change in working hours).

The Training Philosophy

This overview is not to imply that the process team is easily implemented. Real training is required for all employees within the organization. Training only those who are felt to be prime assignees for a team leaves other employees with a

feeling of being left out of an important organizational un-
dertaking. Further, a restrictive attitude on the part of
upper-level management in selection of personnel to be
trained will leave valuable experience behind. Personal interaction methods, problem-solving techniques, even quality assurance practices will become important tools to a
team member. The formal team training each engineer re-
ceives is a motivating force that instills confidence. Team
training provides each engineer with the tools necessary to
improve his communication techniques. Further, the training enables a demonstration of leadership abilities within
the classroom, and provides an atmosphere for improving
personal dynamics.

The Engineering Process Team Makeup

The engineering process team has been held to have a life of
its own. The team is a task-oriented, problem-solving group
consisting of those engineers who are members, as well as
the engineer who is the team leader. Assisting the team at
meetings is another person (not necessarily an engineer)
who can facilitate the operation of the group and encourage
the group toward accomplishment.

Members

The persons assigned to a particular team can be selected
from any area of endeavor pertinent to the problem that the
team will undertake. Usually, those problems that relate
directly to a particular organization are undertaken by
members of that organization. Problems within a particular
department of a large organization (or company) are undertaken by a team consisting of members from many sections
of the department. Problems that affect human resources
are undertaken by teams made up from the total environment of the company and can contain many members un-

der one leader. A large team will require an exceptionally experienced facilitator.

The engineers selected for assignment to an engineering process team are the driving force of the team. Each member is an integral part of each decision and every solution. Members of a team, formally trained in up-to-date management techniques before team assignment, are placed in a position of being able to practice the techniques they have learned.[12] The members select the goals for their team, they present and document the objectives, and they establish the estimated dates of completion for each task or objective. Solutions come only from the members, and the more free-wheeling the team is encouraged to be, the more creative their solutions. It will be found that the success of an engineering process team is in direct proportion to the amount of formal training the individual members received. However, during the course of team operation, members will learn new skills from each other, will learn to work together as a team, and will increase their interrelationship skills. Members will have the opportunity to practice public speaking during their presentations, will become proficient in running meetings, and in general will learn skills that are carried back to the laboratory team for integration.

Team Leader

Those selected to be leaders of engineering process teams are usually elected by the team members at their first or second meeting. They will usually be from the areas of expertise required by the specific team. The skills of the team leader and the team members in the discipline of the process team requirement will be the moving force behind team direction. In some cases a leader will just emerge from the team proper and be accepted by the members. Members of the team may gravitate toward a specific leadership trait shown by a member of the team and that member becomes the natural leader of the team without an election being necessary. However, this is not the usual case. Technical expertise is not the prime requisite for an engineering process team leader and a popular election is usually the most successful manner of selection. That special person must be able to provide encouragement, must have a positive attitude, and must be completely trained in all aspects of process team operation. The team leader must also be able to work closely with the team facilitator.

Team Facilitator

The position of process team facilitator can be a full-time position in the company or it can be given to an employee who has been trained to perform the facilitator function as a collateral duty. It is not unusual for this person to facilitate more than one team.

The facilitator for a team must be at every meeting of that team and must be readily available to speak with team members during the course of each day.

The facilitator should be an expert in company policy. His or her knowledge of the manner in which the company does business enables the facilitator to maintain orderly direction of each meeting, and instills confidence in team members that they are moving in the correct direction. However, one important aspect of the facilitator's duties can not be overlooked. The facilitator must be able to maintain decorum in a meeting when members become excited or vociferous to the point of disruption. When the calm of a meeting erupts into something unproductive, the facilitator is required to take charge and restore calm.

Since team rules usually are established by the senior steering committee, the facilitator must be familiar with all facets of senior steering committee operation.

As with the team leader, a positive attitude demonstrated by the facilitator is a necessity. He or she acts as a coach and a referee, and assists members of the team individually and collectively in their efforts. This special person is responsible for keeping the team on track toward the team goal through objectives, and yet must remain in the background at every meeting.

The Steering Committee

The senior management of a company that makes the decision to convert to a team philosophy takes on a far-reaching task. Although the amount of budget committed is significant, the amount of manpower to be expended in both training and team operating hours is enormous. Senior management accepts the role of steering committee for team operations and is responsible for the development of the policy that the teams will follow. The steering committee decides on the size of the process team program, maintains a monthly meeting schedule with the director of the program, and provides the backing that those in charge of the various teams will need.

THE CODE OF CONDUCT

Although the steering committee is responsible for the rules by which the teams are formed and maintained, the team members, including the leader and facilitator, decide on the rules of conduct to which the engineers will abide.

Common sense and courtesy must prevail in developing the team code of conduct. The following are areas a team might consider as a part of that code:

1. Attend all team meetings
2. Avoid being overly critical
3. Participate actively in brainstorming
4. Give thought outside of meetings to fishbone analysis of a problem (discussed in Chap. 6)
5. Be a responsible member
6. Respect all opinions

5

THE ENGINEERING PROCESS TEAM APPROACH

TEAM STRATEGY

As stated earlier, an engineering process team is a task-oriented problem-solving group with specific objectives leading the team toward the ultimate goal of achievement. The strategy to produce a successful team can be broken down into five steps.

The Aspects

First, all potential members of the team must receive formal training. Second, members are carefully selected. Third, management provides a place for the team to meet comfortably. Fourth, the required meeting time is cleared by senior management through the team members' supervisor or manager. Fifth, the purpose of the team is clearly spelled out and written down by the person who formed the team.

The Purpose

In the case of an engineering team, a statement of primary (generic) purpose might read as follows:

> The team will strive toward the ultimate solution of a problem that hinders the effectiveness of the Company. The Team will incorporate the ideas and energy of each team member in developing the solution, so that each organization in the Company is represented, and through that representation becomes more productive and competitive. Team members are required to provide an atmosphere of encouragement and participation through open and honest communication with one another.

The Working Parameters

In problem solving, the engineering process team works within the scope of three major parameters:

1. **Process selection.** The team first lists the processes of the organization and, through the use of various problem-selection techniques, selects the process to be worked on. Brainstorming is one of the more popular techniques for making selections and will be discussed later. However, it is well to recognize early on that brainstorming is only as effective as the data with which the ideas of the session are formulated. Data collection must not be overlooked as the mainstay of selection and solution.
2. **Problem analysis.** Many techniques for Problem Cause Analysis are available for use by the engineering team. They enhance the ability of the team to make an ultimate selection. Cause and Effect Analysis performed through use of the Fishbone diagram is used extensively today and will therefore be discussed.
3. **Implementing a solution.** The area in which teams

can find much difficulty is in implementing a solution to the problem. Barriers such as financial constraints, unavailability of the various personnel required, and a shortage of time to accomplish the solution can present problems within the implementation spectrum. Many of the problems of implementation can be overcome through running trial implementation scenarios. Once several trials have proven successful, the ideas can be firmly implemented and tracking of the process can begin.

PROBLEM IDENTIFICATION

In the general approach to problem solving, a basic step is identification of the process that has a problem. In the case of the engineering process team, the members consider the process in question (the process that is subject to the specific problem) and use the tools (knowledge) they have been given during the formal training stages. A team in search of a problem of general scope to tackle will find that a free-wheeling brainstorming session will result in a long list from which a problem can be selected. An experienced team focused upon improving a specific process will find the Process Flow Diagram method to be beneficial.

THE BRAINSTORMING SESSION

As a general rule, quantity, not quality, is the basis behind brainstorming. The task of a brainstorming session is to generate as many ideas as possible in the least time. Ideas are not evaluated during the session, and every idea is treated as the one of greatest importance.

Of all the techniques employed for the various phases of problem solving, brainstorming is the most enjoyable. Team members can say whatever comes to mind without censure or analysis.

Getting Started

Prior to beginning a brainstorming session, the team leader or facilitator writes the process and the problem being brainstormed on a large sheet of paper or an erasable board. The leader then establishes that each team member agrees with the problem as written. If necessary, adjustment is made until there is complete agreement among all team members.

Two kinds of brainstorming are employed by engineering process teams during problem identification: structured and unstructured. Each has merit.

The Structured Method

The structured method usually takes place around a table. Each person, in turn, must offer an idea toward selection of a problem. No one is exempt. Each person is encouraged by the other participants. The session can be just as freewheeling as the members of the team will allow it to be. The structured method is often employed during the early days of a newly formed team.

Structured Session Sinks Suddenly

Trim Tim the team leader was pleased with his team. They had proceeded through the initial stages of team development and were now standing on the brink of diving into their first brainstorming session.

Trim Tim explained, "OK, guys, here's the layout. If we're to reduce the amount of time that it takes an engineering change order to move through the department, then we're going to have to have a really freewheelin' brainstorming session. You all know the rules. You've all been trained. Pete, whaddya think?"

Pensive Pete was the least aggressive of the group; however, he was a bright employee with a good track record. Pete stood and said, "Well, one thing I think for sure: there are too many signatures on the thing."

"What are you talking about?" yelled Abusive Al. "Are you nuts? Remove a check signature and the whole system goes to pot."

Pensive took a short step back and began quietly, "Well, I just thought . . ."

"No," hollered Al, "you didn't think. That's the problem here."

"Now, Al," began Trim Tim, "everyone's opinion is impor . . ."

"No it's not," yelled Al, "not if it doesn't make sense."

Meek Mary interrupted. "Maybe we don't all agree that it doesn't make sense, Al," she said in a breathy voice.

"Well," Al said, his jaw jutting out, "I think it's stupid."

With that, Mary and Pete got up from the table, made their apologies for not being able to add anything to the meeting, and left.

The moral of this story is plain:
FREEWHEELING BRAINSTORMING MUST NOT
INCLUDE FREEWHEELING CRITICISM.

With this structured method, bashful members are encouraged to speak out before they have the opportunity to back out. Note, though, that there is a certain amount of pressure to participate, and an alert facilitator is needed to keep the session from becoming intense.

The Unstructured Method

The second method of brainstorming, which is highly results-oriented, is the unstructured method. As the saying goes, "There is good news and bad news." The good news is that the relaxed atmosphere developed by team members' being able to speak out freely whenever they wish provides a great amount of input in a short time. The bad news is that the session is often taken over by the most vocal members and the less assertive members are never heard. When this occurs, the Nominal Group Technique of problem selection (see below) may be employed. Once again, an alert team leader or facilitator plays an important part in ensuring that each member provides input.

Carrying Through

As ideas are generated from either method, the team leader writes each one beneath the agreed-upon problem, and numbers the idea. No adjustment of what has been written is allowed from this point on; the leader writes down just what was heard and then goes to the next idea. During the session, no criticism is permitted, each idea is heard to conclusion, and every idea is recorded for consideration. As long as ideas are being generated, the session remains active. As the session continues, ideas will begin to come more slowly. When ideas cease coming, the session is ended and the ideas are held for consideration.

After the Session

It is unnecessary to begin consideration of the ideas in the same meeting in which the ideas were generated. It is often good to leave the consideration of ideas for the next session, after members have had an opportunity to develop their thoughts outside the group.

The team leader should ensure that the minutes of each meeting are distributed to the team members soon after each meeting. The minutes of a brainstorming session must include a written description of the process that was brainstormed and the ideas for selection of the problem to be worked on. A cover letter accompanying the minutes should instruct each team member to select the five most appealing ideas, list them in priority from one to five, and bring the list to the next meeting.

The importance of brainstorming a problem cannot be overstressed. Brainstorming works. When collected data verifies that a problem actually exists, that problem can usually be solved by an engineering team if enough brainstorming is done. When a brainstorming effort fails to uncover a definitive solution to a problem, more data on the problem must be collected and the brainstorming resumed.

A side effect of the effort is that the people solving the problem become close associates in a common cause and develop a sense of ownership in the problem solution.

Brainstorming Session Number Two

The next meeting begins with the team leader's listing the ideas from the last meeting. The top five selections of each team member, ranked in order of importance, are noted on the board. This voting technique is sometimes referred to as "multi-voting." The votes for each idea are tallied, and the idea receiving the most votes earns the top priority rating for action. The team has arrived at a consensus of opinion, and the problem is ready for analysis.

THE NOMINAL GROUP TECHNIQUE

When the unstructured method of brainstorming is used by an engineering team, the members with the loudest voices often play a far more important role in the selection of a problem than those who are not so loud. The quieter members find no opportunity to present ideas they have brought to the meeting, and therefore find frustration in having done work in vain. Accordingly, the problem has actually been selected by a loud few rather than by the consensus being sought.

A technique found to be successful in correcting this condition is known as the nominal group technique, usually referred to as simply the NGT. In the NGT, each member of the group is solicited by the team leader to write out the problem he wants selected, and to write the reasons for that selection. Those who care to discuss the problems openly rather than write down the problem are welcome to carry on their discussion with the team leader or the facilitator. At the end of the discussion period, the leader collects the written problems and transfers these, as well as those discussed by the speakers, to the display board.

The Advantages

The procedure of displaying all written statements and those obtained in the discussion period provides an opportunity for the team members to hold a general discussion on all problems. Similar problems can be combined into a single descriptive statement. Duplicates can also be combined.

The Method

Each statement on the board is then identified with a letter, and those letters are written on paper by each team member. The members then vote by placing a number beside their first choice corresponding to the number of problem statements. For example, if three statements re being voted on, then the first choice of each team member would be assigned 3 points, the second choice 2 points, and the last choice 1. The team leader then records the point totals beside the problem statements on the board. The problem with the highest total becomes the most important and is selected to be worked on first.

Step-by-Step Problem Selection

1. The team leader writes each problem statement on the board. Examples:
    (1) No criteria for prioritization.
    (2) Budgets are micromanaged.
    (3) Engineering has no voice in new product development.
2. Team members rank their choices of problem in the following manner:

Problem no.	Rank	Points
3	1st	3
1	2nd	2
2	3rd	1

3. The team leader tallies the votes for the problem numbers selected as follows (votes × point value of rank = total points):

Problem 1	Problem 2	Problem 3
3 × 1st = 9	2 × 1st = 6	5 × 1st = 15
4 × 2nd = 8	4 × 2nd = 8	2 × 2nd = 4
3 × 3rd = 3	4 × 3rd = 4	3 × 3rd = 3

4. The team leader adds the totals for first-, second, and third-place selection. The problem with the most points is accepted as the problem to be worked on by the team. The problem with the next-highest number of points is held in abeyance as an alternate. Team agreement is brought about quickly. In the above example, problem 3 won the most points:

$$22 = (5 \times 3) + (2 \times 2) + (3 \times 1)$$

THE PROCESS FLOW DIAGRAM

The process flow diagram is as orderly an approach to problem selection as brainstorming is diverse. Since most engineers used flow charts in engineering school to track a process, the process flow diagram is especially appealing to an engineering process team. Through the use of flow charts the actual path that a process follows can be identified and tracked pictorially (see Figure 6.1).

Advantages

Deviations in the process can be immediately identified through observation and the relationship of the various steps can be compared. One major advantage of the process flow diagram to the engineering process team is that the chart symbols are familiar to most engineers.

Constructing a Flow Diagram

The five steps to constructing a process flow diagram are:

1. The process team ensures a complete understanding of the problem as described in the statement of the problem.

2. A description is written of each function to be represented on the diagram.
3. A template is selected containing the symbols and chart steps needed for the particular diagram being drawn.
4. The diagram is drawn in conjunction with the requirements defined by the team.
5. The function descriptions are written into the diagram symbols (see Figure 6.1).

Flow Diagram Application

The process that the flow diagram represents is examined by the team members. Problems and improvements are noted and areas of potential trouble are discussed.

The flow chart is an excellent device for testing the movements of tracking devices, such as engineering action requests or work orders. Areas that cause slowdowns can be magnified and studied for correction.

To obtain optimal results from the process flow diagram, the team developing the diagram should be the most knowledgeable people available for the process being diagrammed. Two charts are constructed by the team: one depicting the process as it exists, the other in a form that the team considers optimal. The two versions are then compared. Those areas where the two diagrams diverge are the areas to be studied by the team, and are identified as problems.

FINAL PROBLEM SELECTION

Before the team begins a solution mode, the problem selected must be examined critically to ascertain whether a solution to the problem is possible by this particular team. A general discussion is held, and every member is given an opportunity to state his opinion about whether the team can solve the problem. There may be reason at this time to redefine the problem or the scope of the problem. There is, for instance, the possibility that a problem is larger than the team's available resources or their talent to solve it.

When, after careful and final consideration, the team agrees to work on the problem, the solution mode is entered.

6

DEVELOPMENT OF THE PROBLEM SOLUTION

SOLUTION REQUIREMENTS

In both the general and engineering approaches to problem solution, the same rules are applicable. The problem must be identified to the process team by using the various tools available. The problem must be clearly stated by consensus to ensure that each member of the team understands it and agrees with the premise that the problem has a solution within their authority. The problem must be broken into small segments in order to provide areas for solution that are assignable to various members of the engineering process team. When these conditions are met, the team can begin to approach problem solution by examining the cause of the problem, and the effect the problem has had on the process being investigated.

THE FLOW DIAGRAM AS A TOOL

At this point, the team develops a Process Flow Diagram that reflects the latest information the team has acquired. Using this diagram a comparison of the effects each step of the process has on all other steps can be made (see Figure 6.1). An updated Flow Diagram must be maintained through the problem solution mode of activity.

Figure 6.1   Process Flow Diagram.

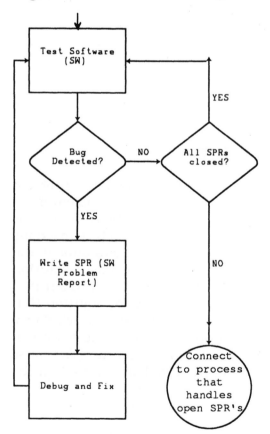

PROBLEM CAUSE AND EFFECT

The challenge of a Total Engineering Quality Management team is to bring into control those processes within a given organization that are out of control. As difficult as it may seem to remain focused on process control while examining the many problems associated with TEQM, that appears to be the one element on which most experts agree. Once a process is brought into a controlled state, then improvements in the process can begin and continuous process improvement can be implemented.

Data Collection and Analysis

To examine the results of a specific process and to recognize that the process is not accomplishing those things on which the organization depends is far more simple than to correct the problem. In many cases, the engineer in charge has not sorted out the facts involved and hence cannot provide the information needed to bring about a solution. In other words, information is lacking. Sufficient data has not been collected to provide a solution.

Check Sheet Method

A method that has been found efficient for data collection in the engineering laboratory is the check sheet. This method of gathering data ensures that the agreed-upon data for problem solution will be collected. The check sheet must contain information on the number of times a specific event occurs. When completed, the sheet will take the physical event of an out-of-control task and turn the collected data into a singular fact that can lead to a solution of the problem (see Figure 6.2).

Figure 6.2   Check sheet (rough).

	Software Quality Assurance				
S/W Lab Errors	Tallies - Week 1				
	Mon	Tue	Wed	Thu	Fri
Variable Definition	11	111	111	1	1111
Comment Spelling	1	11	1	111	11
Coding Errors	1111	1111	111	1111	111
PDL Errors	1	11	11	1	11
PDL/Code Mismatch	1		1		
Other	1111	111	1111	11	111

Event Tracking

In deciding what an individual check sheet will look like, the first step is for the engineering process team to agree on just what events are to be tracked on the sheet. Never losing sight of the advantage of the team members' feeling ownership in the ultimate solution, it is once again important that all team members agree on exactly what is to be looked for. The team leader should make a rough drawing of the check sheet form on the blackboard as the data begins to accumulate (see Figure 6.2).

The group must also decide on the time period during which the data is to be collected. The overall time period divided into manageable data-collection times is the basis for the sheet. The team must keep in mind that the answer to a successful data-collection scheme lies in a clear and easily understood check sheet. Each of the areas into which data is to be entered must be labeled clearly. The open areas of the sheet must contain enough space in which to write understandable data.

Check Sheet Considerations

The team must ensure that the data-collection process is such that personnel are available at the data-collection times and places, and that the data is collected in each time period defined. The data-collection sheet shown in Figure 6.2 is only one example of a blank check sheet. Sample observations laid out in this manner will provide immediate perception of all points that have begun to go out of control.

DATA INTERPRETATION

An important aspect of the data-collection effort is the idea that it is more efficient to collect more than enough data the first time around than it is to attempt to begin an additional effort when the collected data proves inadequate.

Understanding Data

The check sheet method of data collection provides an abundance of data in an orderly form. Although the check sheet could well be replaced by a direct input of data into a computer, thereby removing any chance of human error in computation, most of the applications that concern an engineering process team do not require that degree of sophistication. Once again, simplicity in the approach to SPC is far better than oversophistication.

Once the data for a process has been collected long enough to reveal a pattern to the process, the data must be transferred to another medium for interpretation. An accurate interpretation can then be made and used for correction of whatever process problems have been pointed out.

The Pareto Chart

Understanding the data collected is made simpler through the use of the Pareto chart, a ranking tool used for graphic display of the largest problems compared to all other problems found. The check sheet provides groups of information from which the Pareto chart is constructed, and that information makes up the vertical bars of the chart. Pareto analysis compares the problems found and then displays them in order of prominence. The problem that occurs most often as data logged on the check sheet is shown as the longest bar, standing to the left of all others (see Figure 6.3).

The problem noted less often than any other is represented by the shortest bar, on the far right. Between those two bars are bars representing all other problems in direct proportion to their number of occurrences. The Pareto Chart is an orderly version of the vertical bar chart.

The Pareto Display

Problems shown to exist by the collected data are graphically shown in order of magnitude by the Pareto display. Problems with the most occurrences are pointed out.

As additional charts are constructed during the control process, the process is displayed and tracked until the relative importance of all contributing factors is seen and analyzed. The Pareto chart, in this manner, continually portrays the magnitude of all problems being tracked and lends ease to prioritization. The use of Pareto analysis by an engineering process team will set up process milestones of conditions as they existed before and after the data-collection period. Study of the before and after variations enables accurate judgment of actual progress of corrective measures.[10]

Pareto Construction

The well-constructed Pareto chart gives an accurate interpretation of the problem at a single glance. All pertinent

information and designations should be included in conspicuous areas of the chart. All signs and designations
should be well written and easily understood. Those responsible for the chart development and update should maintain an open-minded approach to its construction and not
become bogged down in large numbers. Events that occur
most often or cost the most are not necessarily the most
important events. One egg hatched will require more immediate attention than 50 eggs in a cold incubator.

A Step-by-Step Method

In the steps leading up to Pareto chart construction, data
must be collected to a point at which the engineering process team is convinced that all aspects of the process problem have been ascertained. The time consumed in data
collection will be in direct proportion to the problem's complexity. The team should ensure that any laboratory historical data on the process is included in their examination.
Once the team members have discussed all data, they can
proceed with construction of the Pareto chart, in seven
steps as follows:

1. Decide what parameters each of the two chart axes
   will represent. For example, consider the number of
   coding errors found by the Software Quality assurance (SQA) department over a period of various time
   segments. Coding Errors Found (CEF) would be displayed on the vertical axis, and the various time segments, such as hours or days, would be displayed on
   the horizontal axis. See Figure 6.3.
2. Chart the errors. Place the various problems uncov-
   ered during data collection into categories (see Table
   6.1). For example: (a) Variable Definitions, (b) Comment Spelling Errors, (c) Code/Preliminary Design
   Language (PDL) mismatch, and so on, until each of
   the problems under consideration can be represented
   as an individual count within some category to be
   displayed.

Figure 6.3 Pareto chart.

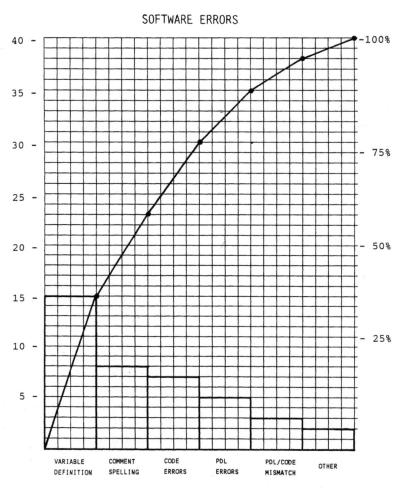

SOFTWARE ERRORS

3. Compare each category to the others by number of occurrences.
4. Prepare a Pareto chart with scales as shown in Figure 6.2. In this case, frequency of occurrence will be on the *y*-axis and software errors will be represented on

Table 6.1 Data for construction
of Pareto chart

Errors	Count
Other	2
Code errors	7
PDL/Code mismatch	3
PDL errors	5
Comment spelling	8
Variable definitions	15

the x-axis. Cumulative percentages will be represented on the right edge of the chart.

5. Represent each category by a Pareto chart bar that is directly proportional to the number of items included. For example, if the number of incorrect variable definitions is 15 and the number of comment spelling errors is 8, then the vertical bars on the Pareto chart would stretch along the y-axis to the correct count and would be directly over the proper designator on the x-axis. See Figure 6.3.

6. After all categories are represented on the Pareto chart, rearrange them with the longest category bar to the left, and follow in descending order until the smallest is on the extreme right.

7. Complete a line that graphically displays what percentage of the total is represented by each bar. This is accomplished by first setting up a chronological display of the data as shown in Table 6.2, to be used as follows.

   a. Add the number for each category to the next number in the row and construct a table as shown in Table 6.3. For example: Start with the largest category (Variable Definitions) and add the next (Comment Spelling), i.e., $(15 + 8 = 23)$. Place the sum on the chart as shown.

Table 6.2 Chronological Arrangement
of Pareto Data

Errors	Count
Variable definitions	15
Comment spelling	8
Code errors	7
PDL errors	5
PDL/Code mismatch	3
Other	2

Table 6.3 Pareto Data Totals and Percentages

Error type	No.	Cum	Cum %
Variable definitions	15	15	38%
Comment spelling	8	23	58%
Code errors	7	30	75%
PDL errors	5	35	88%
PDL/Code mismatch	3	38	95%
Other	2	40	100%

b. Add the sum to the next largest category: Variable Definition + Comment Spelling (23) + Code Errors (7) = 30. Place this sum on the chart as shown and continue until all categories are incorporated into the table. The final total in the table is 40.

c. Obtain the percentages for the table as follows: Percentage is equal to: (cumulative frequency ÷ the total number of occurrences) 100. For example (see Table 6.3):

15 ÷ 40 = 0.38 × 100 = 38%

(15 + 8) ÷ 40 = 0.58 × 100 = 58%

(15 + 8 + 7) ÷ 40 = 0.75 × 100 = 75%

When all additions and divisions are accomplished, the last percentage found should equal 100.

d. Draw the line representing the cum percentage as shown in Figure 6.3. The drawing of the line can be facilitated by ensuring that the graph is of a scale such that all notation is easily discerned and understandable.

Pareto Chart Use

Pareto charts are used for either variable data or attributes data. The chart directly compares the most frequent occurrences of the cause of a problem. The information displayed on the Pareto Chart can be used to develop a plan that reduces the largest category to an additional Pareto chart. In other words, the largest category (incorrect variables) can be broken into various subcategories and laid out on an additional Pareto chart. This method can be followed with each chart produced until the initial problem is reduced to a solvable size and the problem is assigned as an action item within the engineering process team. The solvable size will vary with the ability of the team. As the team works with the longest bar of each Pareto chart, an agreement can be obtained about when is the best time to begin the solution mode. This method can be employed until the team can approach the solution mode with confidence. The major cause of a problem will usually fall out of Pareto construction techniques and produce the need for an even more detailed analysis. When the team members find themselves in that position, the recommended solution method becomes the "fishbone" diagram.

The Fishbone Diagram

A major tool used for cause and effect analysis is the fishbone diagram, so named because of the appearance of the finished diagram.

The cause and effect effort begins only after a problem has been defined and agreed upon by the entire process team. A clear understanding of the problem is important to facilitate breaking a large problem into small, workable pieces.

Engineering Team Goes for the Gold

Lively Len was an excellent team leader. He had led this team of senior engineers through some tough discussions and had been able to maintain decorum throughout. Action items of previous meetings had required the members to collect data supporting their ideas and this meeting would end after the action problem had been identified.

He stood before them about to explain where they were in their pursuits. "The meeting will come to order," he said. The room became silent immediately as each engineer sat forward expectantly on his chair and gave Lively his full attention. "OK then," he said, "what's our problem? If all is well, why do we continue to overrun our budgets and slip schedules?"

Junior John spoke up. "I think it's time to examine the thirty or so items that we developed in our previous meetings as being problems with the department."

Lively smacked his hands together. "Exactly," he said, "I want each of you to take out your list of those problems and weight your choices 1 through 5 and we'll go from there."

It was interesting to note that the top five problems with department budget and schedule were arrived at quickly by this method. The team was ready to proceed to the next step: Development of a "fishbone" cause and effect analysis for the top choice. They developed a problem statement from the five top problem choices written on the board.

> The team stated the problem as follows: "The Engineering Department does not adhere to engineering methodology," and began the chore of coming up with solutions to that problem.

The Dispersion Analysis Fishbone Diagram

Due to the capability of displaying even minute causes of a problem or potential problem, the dispersion analysis type of fishbone diagram has become synonymous with the term cause and effect analysis.[13] This type of diagram graphically represents each cause of some particular effect in the form of small bones (the causes) directing attention to the head of the fish (the problem/effect). Although there are two types of fishbone diagram commonly used in problem solving, this review will deal only with the dispersion analysis type of cause and effect effort. Although the problem classification type of diagram is a viable alternative, however it will not be covered here. Briefly, in the problem classification type of effort, the steps of a process are given more attention and figure more prominently in chart construction.

Using Categories

The fishbone diagram assists the engineering process team in placing into categories possible causes of the problem under analysis. By placing the causes into understandable categories, construction of the chart becomes an orderly process of laying out groups within each category. However, it is important that the team remain aware that a fishbone diagram is a portrayal of the perceptions of the team and that those perceptions are not always factual. Decisions made in a cause and effect analysis session must sooner or later be supported by collecting data. Perceptions must be proven prior to acceptance.

The cause and effect diagram is used by process teams

with a need to explore, identify, and graphically display all possible causes of a specified problem. As the team proceeds further into the exploration, the layout takes on the appearance of a fish skeleton, with the effect of the problem statement placed in the approximate location of the fish's head. The causes of the effect are placed just beneath the dorsal fin. Once the diagram is complete (and a diagram is not easily completed), the relationship between the effect and all possible causes is well represented (see Figure 6.4).

Two Sets of Basic Causes

Many process teams enjoy using one of two sets of basic causes. The selected set then becomes the four major bones

Figure 6.4 Dispersion type of cause and effect diagram.

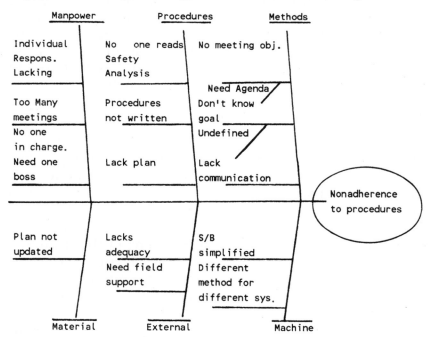

of the fishbone diagram. In technical investigations the four major bones are individually designated manpower, machinery, method, and materials, and are labeled as such. Under these four major categories, many smaller bones are located and attached. In the case of administrative or financial explorations, a good place to begin is to lay in the categories of policies, procedures, people, and plant (facility). A team is not limited to these categories; however, they can be an excellent place to begin.

Once the list of probable causes is completed by the process team, the apparent serious causes are selected and subjected to closer scrutiny. The process team may wish to develop additional fishbone diagrams for the major causes. The path to follow for success suggests looking for those causes that have caused a variation from the normal pattern of things. The team must be certain that the efforts are directed toward curing the cause and not just the symptoms of a problem. The cause and effect analysis will point out differences in the way each engineer understands the process in question and the process problem in particular. Through discussions during chart construction, a better understanding of the problem is realized by each participant.

Problem Identity

Arriving at a common agreement and understanding of the problem is a requirement for success. Ultimate problem solution comes from first identifying the problem fully and completely. When the team can agree that the problem under discussion is the problem to be analyzed, time is taken to ensure that each team member has a complete understanding of the problem. The method to be used in bringing about the solution becomes a part of the discussion. The problem is then broken into small pieces and the major piece is selected for solution.

Construction of the Fishbone Diagram

The process team begins construction of the fishbone diagram as follows:

1. The leader writes a clear definition statement on the board. The definition statement is then discussed and molded by the team until all engineers are in agreement with the statement. The statement is then placed in the area of the fish's head and becomes the problem statement (the problem to be solved).
2. The team determines the major groupings under which the various causes will be listed. Start with either the four Ms (manpower, machinery, materials, method) or the four Ps (policy, people, procedure, process) and make adjustments as required from the input of causes.
3. Data collected by the team in process analysis is introduced into the chart. Causes taken from the check sheets of each team member are listed on the chart either as a major branch of one of the four categories or as an interrelated cause attached like a twig to a main branch. Exhaust the check sheet and collected data.
4. Employ the nominal group technique to ensure that each member of the group has equal opportunity to be heard and to supply input to the chart.
5. Introduce an unstructured brainstorming session and encourage a freewheeling atmosphere to bring about the maximum number of causes. No cause is to be analyzed in this phase. Every cause is immediately written into the chart by the team leader without question.

Selection of Important Causes

As the diagram begins to take shape, an agreement will emerge as to possible corrections for the problem. This will be the outcome of continuously looking for those items that show up repeatedly as bones in the fish. The frequency with which the various causes appear will designate those causes as the real concern. When a conclusion to the diagram draws closer, the need to continually ask "Why does each cause exist?" is very important. When answers that justify the causes are brought out, data must be presented that support the selection. The causes and the data must be questioned until there is a consensus by the team members on the manner of solution.

One way to maintain order within the diagram is for the teamleader to draw a box around those causes that the team has quantitatively proven. In case a cause is unproven but the team members agree there is a distinct relationship between that cause and the effect, the leader heavily underlines that cause. When there is agreement by the team that a listed cause has no proof or relationship, the leader makes no mark and the cause is treated as unimportant. The better the team understands the problem, the more boxes will appear in the diagram.

Ending the Analysis

The cause and effect analysis ends when a solution to the effect (problem) is reached. At this point, a presentation to the steering committee can be helpful and often provides input on directions or limitations that the team may face.

Teamwork Takes A Toll on Toil

Dandy Doug knew more about configuration management than the average bear. There had been a time not too long ago when the engineering documentation numbering system for the company had been a laughing matter for any customer auditing the system. Then Doug obtained permission to form a team of engineers to develop a numbering system that customers could buy into.

The team was carefully selected by Doug and the director for whom he worked. They began meeting once each week and they worked hard at solving the problem of documentation control. The whole team came up with the answer, and they agreed that, by having several new fields added to the existing documentation by the Management Information Section, the agreed-upon system could be implemented quickly.

But there was a problem. To implement the improvements would require some 400 hours of reprogramming at the standard burdened labor rate from finance. And to top it off, additional memory for the existing VAX computer system would need to be purchased. This was going to take some bucks.

Doug called on Mastermind Mannie, the CPI Facilitator, who had been an important part of the team meetings, and asked what he would suggest.

"Well, Dandy," began Mannie, "I guess the only thing to do is to put together a darned good presentation for senior staff, and convince them that you guys are right, that the company could profit by the expenditure, and that we have the resources to accomplish the full solution." Mannie knew he had been eloquent. Doug was impressed.

"OK, Mastermind, set it up for me. I'll get the team together and make some plans."

Three weeks passed and the team had put together a full presentation showing the method for implementation. The finance people had also cooperated and the team could show a payback for the needed funds. That would sell it. They approached the presentation with confidence, knowing they had done their homework.

Doug now stood before seven VPs and the president of the company as well as 23 directors and some other interested folks. The members of his team sat off to one side waiting expectantly for the praise they knew would come.

"Good morning," Dandy Doug began, "I'm Dandy Doug Do-a-lot, and I head up the "Fix the CM scheme for Engineering" team. We are here today to tell you what we have accomplished and to ask you to consider expending the funds to implement our plan."

Things became much more quiet upon the mention of expenditure, but Doug continued without so much as a hiccup. By the time he finished the presentation, he had the feeling that the senior staff was convinced of the importance of what had been accomplished.

"And to finish my summary, ladies and gentlemen, I would like you to look at our last chart, which shows a total payback of expenditure in just 22 months." He smiled and took a step to the side. "Any questions?"

There was dead silence for a few seconds and then . . . THE ROOM BURST INTO APPLAUSE!

The VP of Finance congratulated the team. The VP of Engineering congratulated the team. All the VPs were smiling. And finally the president said quietly, "I want to approve this on the spot. You men and women have done an outstanding job. Thank you."

The team had finished the job.

Dandy Doug Do-a-lot had done it.

TOWARD IMPLEMENTATION

The construction of a cause and effect diagram is not in itself a problem solution. The diagram provides the team with an intricate analysis of the problem and its causes. The team must now proceed toward implementation of a solution. Consideration should be given to constructing additional Pareto Charts for each of the major causes listed. The Pareto would identify the most important causes of the fishbone diagram by noting such aspects as cost, use, or frequency of occurrence. Using the output from the fish-bone and Pareto efforts, the process team should then plan a simple flow diagram that identifies the number of steps in the process with which the team is dealing. Finally, the team may wish to develop control charts based upon simple statistical control of processes. Those procedures will enable the process team to ascertain areas that remain out of control and to institute corrections in accordance with the cause and effect data.

7

STATISTICAL CONTROL
OF PROCESSES

BASIC STATISTICS

The approach to statistics used in the production of control charts is based on a simple premise: *Everything varies according to a predictable pattern.* Variation in the construction, the steps, and the metrics of a process are all changes that take place over time. As new ideas are introduced by those engineers making the process steps to success, various steps will be modified to incorporate the thinking of the employee. Even the experimental aspects of creating change can lead to a positive innovation in an existing process.

PROBLEMS OF VARIATION

There is, however, a problem introduced by variation that can have a negative effect on control. The problem is one of being unable to predict unwanted or drastic variation; it is compounded by being unable to measure the extent of an out-of-control occurrence. Therefore, the need for accurate measurement of each action within the steps of any process becomes a requirement. That requirement is satisfied by the introduction of Statistical Process Control (SPC).

The primary deterrent within industry to introducing statistical methods for the control of a process has been found to be a general lack of understanding among those personnel required to use these methods. The engineer typically remembers only those long-ago school days when statistical problems took up a good part of the study period; at the time, statistics seemed to be a confusing subject, with complex rules and few practical applications.

STATISTICS AND PROCESS CONTROL

As presented here, SPC uses only basic statistical principles that are easily understood and applied. SPC can tell you when you are doing something right, as well as when you are doing something wrong. Once these basic SPC methods are understood, straightforward implementation can be expected to follow.

THE PRINCIPLES

As with the concept of Total Quality Management, SPC cannot be easily implemented without a cultural change taking place within the organization. Senior management must give full support to the operation. Once a receptive environment is brought about that is conducive to creative thinking and employee participation in organizational management, SPC can be successfully introduced. Statistical *thinking* on the part of employees is far more important to the success of the operation than detailed statistical *knowledge*. The formulae and the calculations required to intricately track a dynamic process are easily implemented. Data for many of the required calculations are available and included in Table 7.2. A knowledge of detailed statistical principles is unnecessary.

Employee participation is the key to implementation. Through employee participation in a team atmosphere, problem-solving tools are put into effect and solutions begin to emerge. As reports on the operations of the employee teams are brought to the attention of senior management through regularly scheduled reviews, the process begins to be reinforced by the procedure. Management gains confidence and is willing to expend funds for the SPC program. Employees gain confidence from senior management input at the reviews. More solid participation of all employees is realized within the program.

POSITIVE ASPECTS AS A RESULT

As solutions to process problems are implemented, line managers have a tendency to reward the involved employees with recognition for their additional efforts and the cultural change begins to take place. Reward is necessary. Without reward and recognition, the program will dwindle from lack of interest by the employees, and then die.

THE CONTROL

The control of a process is brought about by controlling the variation within that process. To understand the cause and effect of process variation, the variation must be measured. Simple statistics provide a means by which variation can be tracked and measured. The statistical concepts are easily understood and the measurements are readily taken.

SAMPLING THE PROCESS

Sensible Cecile Sells Sampling

Sensible Cecile enjoyed the information that she de-
rived from statistics. As an engineer in the software
lab, she tracked her schedules and budgets with simple
charts and had been highly successful.

Now she was manager of the lab and wanted to convince the engineers working for her that statistics
could make their output more accurate, and therefore
more effective. She approached Hesitant Hilda and
smiled as she said, "Hesitant, I think we should begin
tracking every software problem report we generate or
receive."

"What," said Hilda in great surprise, "you have really
got to be kidding!"

"No," said Sensible, "we need to know what's happening for planning purposes."

"Aw, come on, Cecile," chimed in Knowing Ned. "We
don't have time to do that stuff. Besides, there's nothing wrong with our output. The guys in S/W Quality
Assurance OK and monitor the errors."

Cecile frowned. She hadn't expected the opposition
she was getting and realized that there was some validity in what she was being told. "Well then, how do
the rest of you guys feel? Are we too busy to track
SPRs?"

"You bet," from Dim Dennis. "Haven't got the budget
to do it, anyway," from Friendly Freda. "We'd need an
extra head," Knowing Ned added.

Cecile thought for a moment and then said, "OK,
guys, your comments are good, but how does this
sound? We'll pull samples from the total collection of
reports we receive and analyze only the sample. Until

we see something is going wrong, we don't need detailed data anyway. Would you buy that?" Everyone
nodded in agreement.

With the exception of Knowing Ned. "Come on Cecile, all those darned numbers will tell us is that something's wrong. We already know that, just from the
reports."

"But consider this, Ned," Cecile said with a smile.
"Just knowing the number of SPRs generated, in process,and closed over a measured period of time will tell
us how fast we work them off. It also tells us how many
to expect in the future. And that's great for planning
purposes. We might even find that you are working too
hard and deserve some help. What do you think of that,
Ned?"

Ned considered Cecile's words for only a minute,
then with a smile of his own said, "I guess I can buy
that. A little planning never hurt anybody."

Cecile had sold sampling to the group.

Sampling is the statistical procedure of removing a carefully selected number of items from a population and then
making a decision about the population based on the information gained from the sample data taken. In taking samples, the items should be selected randomly to ensure that
only inherent variation is present. That is, every cause of
variation must act on individual items selected in the exact
same manner. When a sample is correctly selected, the total
population is represented by the sample data. The population can then be accepted or rejected on the strength of the
sample taken. Sampling done randomly contains all of the
characteristics of its underlying population and variations
in the process will begin to emerge. Figure 7 in Chapter 8 is
a display of sample voltage readings conducted on one pin
of a development ramp generator printed circuit card. The

samples shown in the example were taken over a period of one full shift and are a representative example of one manner of sampling. The time period over which samples are taken is directly proportional to the accuracy required.

When the manner in which samples are to be taken and the area from which they are to be taken have been decided, the sample data is then placed on an understandable chart. Figure 8.7 shows one kind, an $\overline{X}/\overline{R}$ chart, which is presented for discussion in Chapter 8.

THE PATTERNS OF VARIATION

Variation is a natural occurrence found in all but the most highly refined processes. As measurements are taken of the variation within a process, the engineer taking those measurements will notice a distinct pattern begin to take shape. A pattern occurs anytime multiple occurrences of an action are tracked. For example, a competitive archer is required to fire many arrows at the same target throughout the course of a competition. This repetitive action provides

Table 7.1 Archery competition target data for 20 arrows

Arrow no.	Score	Arrow no.	Score
1	7	11	6
2	7	12	6
3	9	13	10
4	10	14	7
5	7	15	5
6	6	16	7
7	8	17	8
8	4	18	6
9	9	19	7
10	7	20	9

an excellent opportunity to track each shot the archer makes, and tally the results for examination. When the competition is completed the archer's record could look like the data in Table 7.1. The pattern in the table indicates that the ability of this archer is centered around the number 7 on the target. Conclusions can then be drawn as to what the archer can do to perform at a more accurate rate.

It can be said not only that all processes contain variation, but that all variations will evolve into a distinct pattern. As the pattern begins to form, the frequency with which a particular variation occurs within a specific area will show the manner in which all variations are distributed. Therefore, the pattern caused by the variations (or samples) is commonly called a *frequency distribution.*

THE FREQUENCY DISTRIBUTION

To understand a frequency distribution and to use information furnished by it, some knowledge of basic statistical terminology is useful. The language of statistics for process control is taken directly from terms common to arithmetic and elementary algebra.

Each group of samples selected from a population makes up a small part of the population and is referred to as a subgroup. For example, five individual samples taken at 10:00 A.M. on the 8th of June are the 10:00 subgroup. Each subgroup has an average which is calculated from the samples that make up that subgroup.

The Mean

The average of a subgroup is referred to as its mean and represented by the symbol $\overline{X}$. As more and more subgroups are collected in the data-collection phase of the operation, the average of the subgroup averages is recalculated. This composite average figure for all subgroups is called the grand mean and is represented by the symbol $\overline{\overline{X}}$. As samples are taken, it can be noted that in many cases the sample means tend to group around a specific value. This phenomena is referred to as a central tendency.

The mean of any set of numbers can be easily calculated by adding the individual numbers and dividing by the total number (n) of measurements taken.

Two other measurements are important for use in statistical process control: the median and the mode.

The Median

The median is the midvalue of the data collected. Half of the data is always above the median, and the other half is always below the median. For example: the sample 1, 3, 4, 8 has no whole number as the middle value; its median is 3.5. The samples 2, 2, 3, 5, 8 and 3, 3, 3, 3 both have medians of 3.

The Mode

The mode is the value that occurs most frequently when the data is tabulated. It is the most likely sample value selected. Example: The mode for the target scores in Table 7.1 is 7.

Frequency Distribution Demonstrated

To demonstrate a frequency distribution in simplest form, suppose that 10 test engineers are receiving two 5-volt power supplies across their work station each day for five days. Their interest is in measuring the output voltage of each supply and logging the voltage of each unit to the closest full

Figure 7.1 Frequency distribution tally sheet.

OP Voltage	P/S Count	FREQUENCY
1	11	2
2	111	3
3	ꟷꟷꟷ	5
4	ꟷꟷꟷ 111	8
5	ꟷꟷꟷ ꟷꟷꟷ ꟷꟷꟷ	14
6	ꟷꟷꟷ ꟷꟷꟷ ꟷꟷꟷ 111	18
7	ꟷꟷꟷ ꟷꟷꟷ 1	11
8	ꟷꟷꟷ 11	7
9	ꟷꟷꟷ	5
10	11	2

volt. The test manager is interested in displaying the results of one week's measurements.

Figure 7.1 is an example of the tally sheet the manager used to display the raw data. The output voltages for the supplies are displayed in the first column, the count of the supplies with that output voltage is displayed in the second column, and the sum of the counts is displayed in the third column. The data collected by the test manager produces a bell-shaped (normal) curve (see Figure 7.2).

Essential information for statistical tracking of a process is obtained by examination of the data contained within its frequency distribution.

Sample Range and Average

By observing all the samples taken in a process, the range (R) over which the sample values spread and the average value ($\bar{X}$) of the collected samples can be easily computed. The range is a common measure of the dispersion of the data found by simply subtracting the lowest measurement from the highest. In the rare case that maximum–minimum is negative, take the absolute value.

Figure 7.2   Bell-shaped curve.

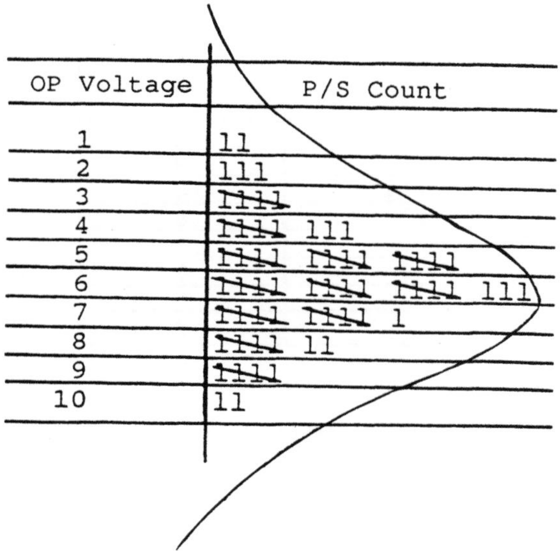

OP Voltage	P/S Count
1	11
2	111
3	卌
4	卌 111
5	卌 卌 卌
6	卌 卌 卌 111
7	卌 卌 1
8	卌 11
9	卌
10	11

Standard Deviation

The accuracy of the plotted data will be in direct proportion to the sample size of the data collected. When an engineering process team collects data, the goal is to determine whether or not the process being measured meets with the engineering specification for that process. When samples are laid out in a manner such that a bell curve is formed, whether normal or distorted, conclusions can be drawn by observation of how far the process being measured spreads from a line drawn through the center of the curve. By comparing the process spread to its specification, variation outside the specification is noted. When the extreme ends of the curve protrude past the established limits, the team has produced a curve that shows quantitatively how many variations exceed the allowable limits. Figure 7.3 is laid out as a pattern of standard deviations. Population standard

Figure 7.3 Standard deviation layout.

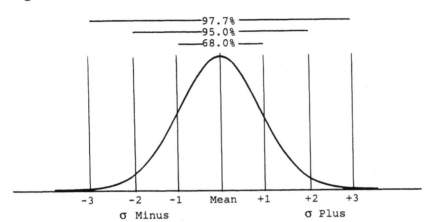

deviation is annotated as σ (sigma) and characterizes the pattern by which the samples cluster around the center of the curve. When the mean is calculated by measuring the entire population, there is no variability attributable to the sampling.

Estimated standard deviation is annotated as σ̂; its formula is shown in Figure 7.5. Its formula is shown in Figure 7.5. It is important to note that this estimated standard deviation is best used when the data under examination is normally distributed and the process is in control; however, σ̂ may be used at any time, and is sometimes found to be a good estimator of variation for nonnormal distributions.

Therefore, a frequency distribution is the curve produced by sample data from a population distribution. Frequency distributions and standard deviations are important parts of Statistical Process Control. However, they provide only a partial look at the available data. Another excellent manner of supplying an understandable glance at the available data is through a histogram. The histogram, drawn in conjunction with the data from which the curve is formed, is simply a frequency distribution laid out in block form. By laying

Figure 7.4   Typical bell shape of a histogram.

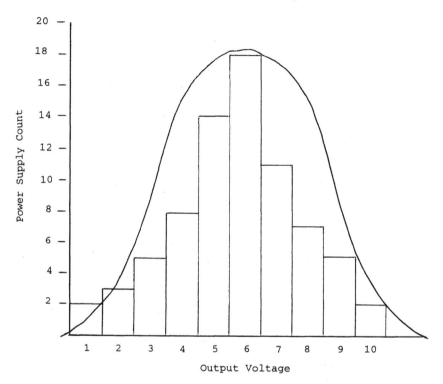

out the histogram, a decision can be made about whether the process being examined is producing a bell shape. A bell shape indicates that the process is in control.

The histogram in Figure 7.4 was drawn from the power supply data gathered by the test manager for Figures 7.1 and 7.2. The process depicted is normal (bell-shaped) and stable. Were the figure to show other than a bell shape, the conclusion could be drawn that the process was not normal, that the variations depicted were due to assignable causes (causes other than those created "by chance"), and that those assignable causes were influencing the shape of the histogram.

Figure 7.5 Formula for calculating estimated sigma.

$$\hat{\sigma} = \bar{R}/d_2$$

Note: d_2 is a constant that is dependent on the size of the sample. Values for d_2 are found in Table 7.1.

When More Accuracy Is Required

Samples can be collected at specific clock intervals, thereby introducing time-dependent data into a control chart. The chart should display this information over as long a period as is required to fully show the trends of the process under examination.

Standard Deviation Estimated

The population standard deviation is often unknown or difficult to measure (e.g., costly or time-consuming). Much time could be spent just in making the measurements. For the purpose of SPC, an estimate of the population standard deviation is usually used. An estimate with desirable properties can be made when the population that is being examined is normal and stable. This estimate is denoted by $\hat{\sigma}$ and can be calculated easily through use of $\bar{R}$ (the average of the sample ranges). An example of the calculation method is shown in Figure 7.5.

The Normal Distribution

A bell-shaped distribution curve is referred to as the normal (or Gaussian) distribution curve. In general, the normal

Table 7.2   Table of statistical constants

Tallies in subgroup	Constants for $\bar{X}$ charts		Divisors for estimate of std. dev.	Range lower control limit	Range upper control limit
n	A	A_2	d_2	d_1	d_4
2	2.12	1.880	1.128	—	3.267
3	1.73	1.023	1.693	—	2.574
4	1.50	0.729	2.059	—	2.282
5	1.34	0.577	2.326	—	2.114
6	1.22	0.483	2.534	—	2.004
7	1.13	0.419	2.704	0.076	1.924
8	1.06	0.373	2.847	0.136	1.864
9	1.00	0.337	2.970	0.184	1.816
10	0.95	0.308	3.078	0.223	1.777

curve has its mean equal to both its mode and median. The curve is symmetrical about its center line (mean) and slopes away from the center line to infinity (see Figure 7.4).

A normal distribution is completely specified by using just two measurements: One is its mean ($\bar{X}$) and the other is its standard deviation (σ).

A normal distribution is formed when many samples taken from a process cluster close to the average value of all samples. There is a tendency for fewer samples to appear as the input moves away from the center line (or average value). Conclusions drawn from normal distribution curves can be applied to control variations within the sampling media.

Normal Curve Distortion

Samples taken from any process can be expected to perform in this same manner, just as long as no new causes of variation are introduced within the process or during the sampling. In other words, when the causes of variation come from the process and not an external source, the curve produced can be expected to act in a repeatable and predictable manner. However, when a variation is brought about by some cause other than a process cause, the curve of the data may become distorted and lose its bell shape.

8

PROCESS CONTROL CHARTS

INTRODUCTION

Control charts provide a graphic tracking method whereby measurements and samples taken can be laid out against the time variable for the data-collection period. Through the use of simple charts, the variation of the process under surveillance, and the time and place of those variations, can be noted and corrective action initiated. In the course of tracking data, two distinct types of variation can be found to occur: common cause and assignable cause.

COMMON CAUSE VARIATION

Variation manifested by causes found to be inherent in a process is called common cause variation. Common cause variation can be found in any system or process and demonstrates stability over a long period of time.

ASSIGNABLE CAUSE VARIATION

Variation that occurs due to a significant outside cause is known as assignable cause variation. These variations can be attributed to such occurrences as changing the operator of a specific piece of equipment or introducing a new method of operation.

Careful Clarence Searches Systematically

Careful Clarence worked for many years in the rate-table area before being placed in charge. His methodical step-by-step approach to system test had never been very popular with the guys, but was appreciated and noted by the Director of Engineering. Our scene opens at rate-table #1, just outside Careful's office.

Rad Rod, having just placed a system onto the rate-table: "Doggone, this stuff ought to work like a champ. Those software people said they had only found five bugs in the whole system. I'll bet even old Careful won't find anything wrong with that."

Careful Clarence, sitting at his desk: "Man, five bugs. That's five bugs in just 500 lines of code. Hmmmm." He reached for his calculator. "That's 0.01, a *1% defect rate!* Which is exactly 10 times higher than their average. Guess I'd better go have a chat with those software people. We need to begin charting the defects and find out what's really occurring."

Rad Rod, seeing Clarence walk from his office: "Son of a gun, all that guy does is sit in his office looking at his charts, graphs, and printouts. Why the heck isn't he out here helping us solve problems?"

CHARTS FOR SPC

An efficient method for determining the cause of process variation is the use of control charts.[11] Many types of control charts have been developed to enable data tracking at various levels. Several of these are particularly suited to Statistical Process Control as it relates to the engineering or the production environment. The run chart stands alone and is an excellent tool for doing simple tracking over a long period of time. The histogram gives a snapshot of existing process conditions. Two more charts that will be represented in this text are the $\bar{X}$ chart and the range chart. These two charts are usually maintained in proximity to each other so they can be compared.

The Run Chart

The simplest visual representation of data displayed over a specific time period is the run chart. This chart can be adapted to most requirements and is usually employed when there is a need to display a long-range change in the average variation of the process under consideration. The run chart is not only simple to read and understand, it is also easy to construct: As each measurement is made at some specific interval, that point is plotted on the chart and the time that the sample was taken is written above the entry.

Constructing a Run Chart

To begin construction of a run chart, the engineering process team makes a decision on what the time intervals will be between the measurements being taken. The governing factor in time interval selection is the total period over which the process will be observed.

For example, consider an engineering team formed to investigate the feasibility of shifting from a process of hand application of conformal coating to machine application. The investigation will show whether, prior to mounting of surface-mount devices on integrated circuit cards by the developmental laboratory, the automatic dipping machine being evaluated will apply the coating material in a thickness that is uniform and consistent. The team makes a decision to construct a run chart and thereby ascertain the average thickness of the coating on dipped cards. If the coating operation is found to be acceptable, the team will establish a tolerance and the company will shift to the automatic process.

The machine produces a coated card every 15 minutes, and the team decides to check every other card for four hours, a total of eight cards. To ensure that only inherent variation is present during sample taking, the machine will be put into operation and allowed to operate with no external intervention for the entire four hours. Thus, all variations in coating thickness during the sample period will be to chance causes. No operator adjustments would take place.

In this example, as sampling took place, the team made the measurements and began chart construction.

Run Chart Layout

The data collected by the team produced the chart shown in Figure 8.1. Note that coating thicknesses are laid out along the y-axis and the time periods are tracked along the x-axis. The sample graduations should be laid out such that the expected average of the process will lie somewhere near the middle of the axis.

Variation can easily be seen when the points on the chart are connected by a line. When many samples have been taken over a long period of time and the average is not readily discernible, the formula in Figure 8.2 can be employed.

Figure 8.1 Run chart.

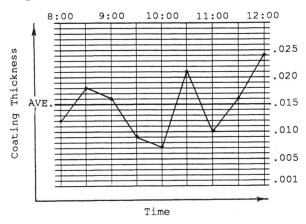

The run chart is a simply produced method for identification of trends or shifts in the average of a process. In the case of a shift in the number of average events, the shift is readily seen as events running on one side of the average line. Display of the data collected in the above example for the number of cards coated and inspected during the course of four hours, and distributed in hourly increments, could well have been shown as a histogram (see Figure 8.3).

Figure 8.2 Formula for taking the average.

$$\overline{X} = \frac{X_1 + X_2 + X_3 + \ldots X_n}{n}$$

where n = number of samples.

The average, designated $\overline{X}$ (in speech, "X-bar") is equal to the sum of all samples divided by the number of samples taken.

Figure 8.3   Comparison of a histogram and a run chart.

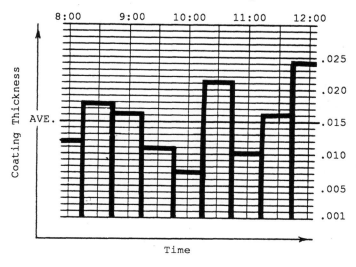

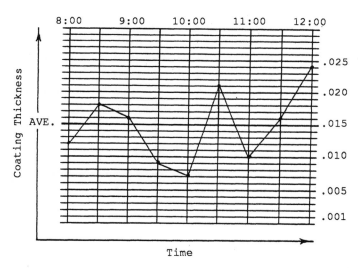

The Histogram

The histogram is laid out in the same manner as the run chart. In the example shown, the coating thickness is on the y-axis and the time of the sample collection is on the x-axis. The difference in the two types of charts is in the manner of connecting the variations. At the top of each hourly period in the example, a line is drawn across the period and a bar constructed from the sides of the line to the x-axis. When this is accomplished for each time period on the chart, a series of side-by-side bars graphically depict the process variation.

Constructing a Histogram

To construct a histogram, the team first ascertains the range of the data with which they are dealing. This is accomplished by subtracting the lowest sample value from the highest value.

Configuration Decision

Decide how many bars will represent the samples taken. When many samples are taken over many periods there may be difficulty in deciding how many periods or classes can be represented by a bar. In these cases, the number of representative bars in the graph should be ascertained by use of the table of data shown in Figure 8.4.

Greater accuracy is produced by a larger number of samples. A general rule is that the number of samples represented by a histogram should not be less than 50.

Determining the Span of a Bar

Determine the span of the period each bar will represent by dividing the range by the number of classes. A general rule is that at least 20 observations should have been made prior to beginning construction (see Figure 8.4).

Figure 8.4 Determining histogram span.

Number of Readings   Number of Intervals

Number of Readings	Number of Intervals
0 to 50	5 to 7
50 to 100	6 to 10
101 to 150	7 to 12
150+	10 to 14

Span=range divided by number of periods

Example: 3.1/10 = .31

Completing the Histogram

Set up a y-axis representing the samples and an x-axis representing the periods. Draw in the bars depicting the data and the histogram is complete.

The $\bar{X}$ and Range (R) Charts

Although many varieties of control charts can be applied to various process situations, the most useful charts for an engineering process are the $\bar{X}$ chart, showing the average operation of a process, and the R chart, which shows the range over which the process operates. $\bar{X}$ and R charts are commonly used in conjunction with each other. They are particularly effective when time and consideration are given to what is to be measured and how the measurements are to be accomplished.

Where to Begin

Data is needed to construct the $\bar{X}$ and R charts, data that comes in the form of many small samples. The sample size will usually be determined by the quality specifications in effect for a particular contract. However, the sample must be large enough to ensure accurate estimates. All charac-

Table 8.1 Random number table

3	8	23	50	46	4	9
6	8	5	32	4	34	16
10	45	12	9	14	43	23
17	5	7	23	43	7	15
45	32	46	12	42	5	18
6	14	8	16	21	5	34

teristics of the population from which the sample was chosen should be present in the sample. Juran[9] suggests that a random number table, such as that shown in Table 8.1, be used. To extract six samples, for example, the table could be entered at line six. The number of elements in each sample would therefore be 6, 14, 8, 16, 21, and 5.

Whatever the sample size required, each sample must be taken so that only inherent variation is present. Inherent variation is that variation due strictly to chance. Since inherent variation is probably a part of the process in use, its elimination would be highly unlikely. Once a decision is made about the sample size and the time period over which the samples are to be collected, the data-collection operation begins. Much is gained by keeping sample sizes to under ten. In fact, a sample size of five is usually adequate for engineering applications. Time periods should also be kept small at the beginning. Periods of a half hour or an hour are commonly adequate at the beginning of process tracking; however, as initial information is gained, the time periods can be lengthened to two hours, or even more. Once again, no matter what time period is used, every sample should be taken from a single source. When each sample is as much alike as possible, the variation present should be inherent variation.

Of the many types of control charts produced, the $\bar{X}$ and R charts are the frontrunners. Figure 8.5 shows a simplified version of both. Note that each measurement shown in the $\bar{X}$ chart could represent a group of daily samples taken over a

Figure 8.5 Simple $\bar{X}$ and R charts.

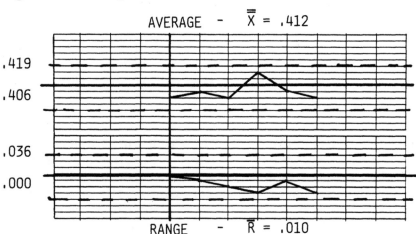

period of five days, or could represent a period of time shown in hours. The important aspect of the data shown by the $\bar{X}$ chart is that each point on the chart is an average value of some subgroup.

Control Limits

The dotted lines above and below the center line ($\bar{\bar{X}}$) are designated the upper and lower control limits. When an $\bar{X}$ chart represents a process that is stable within these control limits, the measurements tend to develop into a normal distribution curve when plotted. An $\bar{X}$ chart shows the average of the measurements for a distinct subgroup represented by a single point on the chart. An R chart, on the other hand, shows the ranges for each subgroup.[12]

Constructing $\bar{X}$ and R Charts

The construction of the $\bar{X}$ and R charts takes place on the same form and therefore will be discussed as a single topic. The same considerations are given to the data for both charts.

Over the long run, control charts will inform the process team as to whether variation in the process is inherent or whether the variation is due to assignable causes. The correction of individually assignable causes will lead to improved quality. The consistent use of $\bar{X}$ and R charts during the course of a process will furnish a useful history for continuous process improvement. A step-by-step method for constructing an X/R chart is described below.

The following procedure assumes that the engineering process team has blank forms of the type available through the American Society for Quality Control for the $\bar{X}$ and R charts (see Figure 8.6). These blank forms contain all the necessary fields for inputting the data, and have areas for the time periods chosen.

The procedure is used to fill in a chart from data collected as shown in Figure 8.7. All calculations will then be based on that data. The data simulates an investigation by a team of design and system engineers into the problem of a suspected intermittent ground on one pin of a ramp generator card. The card is being produced for R&D purposes in the Design and Development laboratory. All measurements are in millivolts to ground. Five consecutive measurements are taken each half hour.

STEP ONE. Post the blank chart in a conspicuous place and ensure that each person involved with the process under examination understands its use.

STEP TWO. Begin taking samples and place the numbers on the chart in the order they are taken. Continue to take samples over the time period of the process. In Figure 8.7, the period is one full working shift from 8:00 A.M. to 4:30 P.M.

STEP THREE. Calculate the average of each sample. That is, sum each sample and enter the value into the SUM row on the chart. As the sample numbers are considered during this process, mark the high and the low in each subgroup for future use in calculations. For example, the

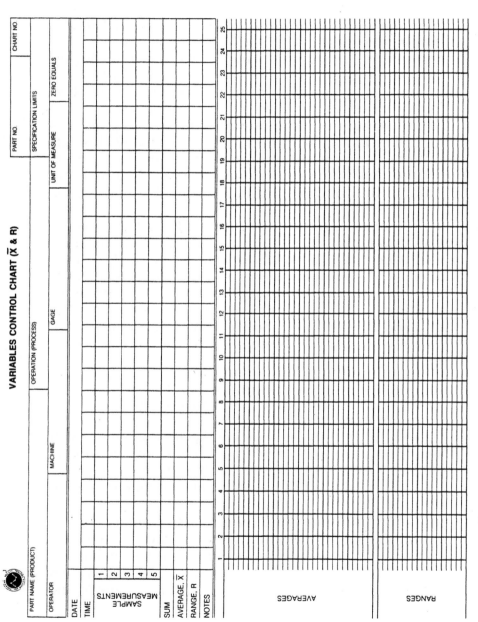

Figure 8.6 $\bar{X}$ and R chart form as constructed by the American Society for Quality Control (ASQC)

igure 8.7 Completed control chart.

VARIABLES CONTROL CHART (X̄ & R)

| ART NAME (PRODUCT) *RAMP GENERATOR CARD Q2* | OPERATION (PROCESS) *ASSY LINE #2* |
| OPERATOR *X X X V* | MACHINE *METER* *SIMPSON OX9* | GAGE *AB-CAK-07* |

ATE																	
ME		8⁰⁰	8³⁰	9⁰⁰	9³⁰	10⁰⁰	10³⁰	11⁰⁰	11³⁰	12³⁰	1⁰⁰	1³⁰	2⁰⁰	2³⁰	3⁰⁰	3³⁰	4⁰⁰	4³⁰
SAMPLE MEASUREMENTS 1	3	6	9	8	7	4	3	2	1	6	4	3	5	7	8	2	6
2	4	3	7	6	4	3	1	1	5	9	8	7	6	5	2	1	7
3	2	4	6	8	9	1	7	3	5	6	8	7	8	9	2	5	4
4	4	6	3	5	8	7	6	4	3	8	7	5	4	3	6	2	1
5	5	4	3	9	8	7	4	8	6	7	4	3	2	8	7	5	4
M	18	23	30	36	36	22	21	18	20	36	31	25	25	31	25	15	22
ERAGE, X̄	3.6	4.6	6.0	7.2	7.2	4.4	4.2	3.6	4.0	7.2	6.2	5.0	5.0	6.2	5.0	3.0	4.4
NGE, R	3	3	6	4	5	4	6	7	5	3	4	4	6	6	4	4	6
TES																	

8:00 A.M. subgroup has a high of 5 and a low of 2, with a total of 18.

STEP FOUR. Divide SUM by the total number of samples taken for any particular time. Place this answer in the row marked "Average, X̄." The sum of 18 divided by the number of samples (5) equals an X̄ of 3.6.

STEP FIVE.   Calculate the overall average (the grand average) by adding all the figures in the "Average, $\bar{X}$" row and dividing that total by the number of readings in the row. Figure 8.7 has a total of 17 readings in the "Average, $\bar{X}$" row. Those readings added together total 86.80, and 86.80 divided by 17 equals 5.11. Therefore, the average of the averages ($\bar{\bar{X}}$) is 5.11

STEP SIX.   Find the range by subtracting the smaller number from the larger number. Record that difference in the row marked "Range, R" in the same column as the figures used for calculation. The 8:00 A.M. range is equal to 5 minus 2, for a difference of 3.

STEP SEVEN.   Calculate the average range ($\bar{R}$) by summing all range entries and dividing by the number of entries. The 17 range readings added together equal 80, and 80 divided by 17 equals 4.71. Therefore, the average of the ranges is 4.71.

Note: At this point, the engineering team has collected the necessary data for developing a graphic plot of the averages ($\bar{X}$) and the ranges (R). The data can be placed on the graph areas of the form to coincide with the times taken.

Before plotting data, set the scale of the charts for the data anticipated.

STEP EIGHT.   To calculate the graph scales, begin by first finding the largest and smallest averages ($\bar{X}$) and the largest and smallest ranges (R). Set the scales for both charts so that the largest and smallest of each fit within the chart area. In the example, the highest average ($\bar{X}$) equals 7.2, the lowest 3.0; the highest range (R) equals 7.0, the lowest 3.0.

STEP NINE.   Plot the data, using the average data for the top graph and the range data for the lower graph, and connect the dots, forming one line for the averages and another for the ranges.

STEP TEN.   Using the figures obtained for the overall average calculated in Step Five and the average range found in

Step Seven, draw heavy lines at those points from one end of each graph to the other and label them.

STEP ELEVEN. Calculate the control limits for each graph in the following manner. For the upper control limit of the range (R):

$$UCL_R = d_4 * \bar{R}$$
$$= 2.114 * 4.71 = 9.96$$

(For d_4 and other constants, refer to Table 7.2, Table of Statistical Constants.)

For the lower control limit of the range:

$$LCL_R = d_3 * \bar{R}$$
$$= 0 * 4.71 = 0$$

For the upper control limit of the average $(\bar{X})$:

$$UCL_{\bar{X}} = \bar{\bar{X}} + (A_2 * \bar{R})$$
$$= 5.11 + (.577 * 4.71) = 7.83$$

$\bar{\bar{X}}$ is the grand average.

A_2 is taken from Table 7.2, and $\bar{R}$ is the average range.

For the lower control limit of the average $(\bar{X})$:

$$LCL_{\bar{X}} = \bar{\bar{X}} - (A_2 * \bar{R})$$
$$= 5.11 - (.577 * 4.71) = 2.39$$

where $\bar{\bar{X}}$ is the grand average, A_2 is taken from Table 7.2, and $\bar{R}$ is the average range.

STEP TWELVE. Using the values obtained for UCL_R, LCL_R, $UCL_{\bar{X}}$, and $LCL_{\bar{X}}$ in Step Eleven, draw heavy dashed lines at those points from one end of each respective graph to the other and label them appropriately.

Control Chart Data Interpretation

When the engineering process team has completed inputting all data into the $\bar{X}/R$ Control Chart, all that remains is

Table 8.2 Data associated with Figure 8.7

X	5.11
$\bar{R}$	4.71
d_3 (5 samples)	0.00
d_4 (5 samples)	2.114
UCL_R	9.96
LCL_R	0.00
$UCL_{\bar{x}}$	7.83
$LCL_{\bar{x}}$	2.39

to interpret the findings and react accordingly. The Team can continue with the established process when the process is found to be in control, or make changes that correct an out-of-control situation.

All Data Points Inside Control Limits

When all data points are found to fall within the control limits for both the range and averages charts, the process is in control. This situation indicates that only inherent variation is present in the process under examination, and therefore the process can be continued as it exists (see Figure 8.7). The team will continue to plot data to ensure that the process remains in control.

There are times when one or two points fall outside the range control limits, and they are clearly rare or erroneous values. Acceptable procedure is to throw out those extreme points and then refigure overall average range (R).

Three Data Points Out of Limits

When three or more data points are found to extend outside either the range control limits or the average control limits, the assumption is usually made that the process is out of statistical control and that inherent variation is unstable. Assignable causes are probably present and should be found and removed. Eliminate those causes and recalculate the grand average ($\overline{\overline{X}}$) and the range average ($\overline{R}$).

New limits must now be calculated. When it is found that data points extend outside the newly calculated limits, the process is out of control and the causes of the data points outside the limits must be found and eliminated. When there are no points out of limits with the new limits calculated, the process should then be carefully checked to ascertain whether or not all assignable causes have been totally eliminated.

Control Chart Summary

No fewer than 50 measurements should be made over the selected time period, usually in groups of five. Those five sample values become a subgroup and are averaged. In other words, a sample value is designated by X and the average of a subgroup of sample values is designated by $\overline{X}$. It follows that all $\overline{X}$ values are then plotted on the chart at the exact time they were taken. The R chart is filled in the same manner. $\overline{X}$ is calculated by simply adding the averages of the subgroups and taking the average of these averages. In this manner, the grand average ($\overline{\overline{X}}$) is computed and placed at the midpoint of the x-axis. Once $\overline{\overline{X}}$ and $\overline{R}$ (the average range) have been established, the upper and lower control limits for both the $\overline{X}$ and the R chart can be calculated by using the following formulae:

$\overline{X}$ Upper Control Limit:

$\quad$ UCL $= \overline{\overline{X}} + (A_2 * \overline{R})$

$\overline{X}$ Lower Control Limit:

$\quad$ LCL $= \overline{\overline{X}} - (A_2 * \overline{R})$

Range Upper Control Limit:

$\quad$ UCL $= d_4 * \overline{R}$

Range Lower Control Limit:

$\quad$ LCL $= d_3 * \overline{R}$

The constants A_2, d_4, and d_3 in the above calculations are taken from Table 7.2. These constants are also found on the back of the control chart blanks furnished by the ASQC and recommended for use in Figure 8.6.

The example shown in Figure 8.7 is a typical layout for an $\bar{X}/R$ comparison. For a typical "in-control" process, the pattern shown on the charts would appear to be totally random with no recognizable pattern. Most points would be near the center line, as shown in the example; however, a few points would approach the control limits.

PROCESS CAPABILITY

All sources of variation must be considered when calculating the capability of the process being examined by the team. There is a distinct possibility that the process, even when in absolute control, cannot provide an output satisfactory to the team. Unless the process is satisfying the customers' requirements, the process is suspect and a change should be considered. The question arises as to just how the engineering process team can make a decision about whether or not the process is capable of doing the required job.

Process Capability Indices

Establishing process capability can be a laborious and tedious task. Much time can be consumed in developing and analyzing data with the mathematical formulae available. However, for the engineering process team, the simple formulae for developing the capability index (C_p) for the process and C_{pk} for two-sided specification limits are just about all the formulae they will need. It is important to recognize that process capability is a consideration of the engineering specifications as well as the control charts. To ensure an accurate capability index, data should be collected over a period of at least one month. The longer the period of data collection and plotting, the more accurate the index will be.

A capability study begins with data collection and the construction of control charts. The process control limits must be carefully calculated and process data plotted on the charts. Ensure that the process is in control before calculating process capability. An index taken from a process that is not in control and not normally distributed will be inaccurate.

Keen Ken Keeps Kool

Jumping John had just finished his first data collection chore. The task had been assigned by Keen Ken, the System Lab manager.

John was unhappy. "Come on, Ken. I don't like to complain about things like this, but I'm a system engineer, not a statistician. I've collected and plotted till I'm blue in the face, and just like I said from the beginning, the process is a good one. Look at that curve." He pointed to a conspicuously posted chart on the lab wall. "We're not only in control, we're normally distributed."

Ken smiled paternally. "Yeah, Jumpin', I thought we'd be within your calculated control limits, but how about this?" He held out a sheaf of papers. "We're not within the spec limits established by the Division. What we have to do now is draw these spec limits on your chart and find out why our efficiency is less than it should be."

Jumping John shook his head dolefully. "Aargggh. What now?"

Keen Ken remained undaunted. "Our next step is to find out whether or not our process can ever meet our spec limits."

> John's interest heightened. "OK. How do we do that?"
>
> Ken smiled again and gave John the news: "We'll now calculate C_p and C_{pk}."

Development of C_p

The capability index C_p portrays the relationship between the upper and lower specification limits and the estimated standard deviation of the process. As previously stated, a set of $\overline{X}$ *and R* charts must be developed to ensure that the process is in control. Once these charts are developed (refer to the sample charts in Figure 8.7), the C_p index is found as follows:

$$C_p = (\text{USL} - \text{LSL}) / (6 * \hat{\sigma})$$

where USL is the upper specification limit and LSL is the lower specification limit.

These two limits are established by a process authority such as the engineering department, the contract requirements, or industry standards. They can be different for each chart developed.

For example, in Figure 8.7, assume an upper specification limit of 7.0 and a lower specification limit of 3.0. Assume the limits were given to the lab manager by the team of engineers investigating the problem.

Estimated Process Standard Deviation

In the case where the process is in control, $\hat{\sigma}$, which is needed as input for the C_p formula, can be estimated from the control chart data in Table 8.2 as follows:

$$\bar{\sigma} = R / d_2$$
$$= 4.71 / 2.326$$
$$= 2.02$$

where $\bar{R}$ is the average of the subgroup ranges and d_2 is a constant value from Table 7.2.

Now, using the calculated $\hat{\sigma}$ in the C_p formula:

$$C_p = (USL - LSL) / (6 * \hat{\sigma})$$
$$= (7-3) / 12.12$$
$$= .33$$

The process capability C_p for the process depicted in the Figure 8.7 chart is .33. A C_p of 1.0 or greater indicates that the process under investigation would be capable of meeting the stated specification limits.

Development of C_{pk}

C_p indicates only how the difference between the upper and lower limits of the specification relate to the capability of the process. However, C_p does not show how well the process average ($\bar{X}$) is centered. The C_{pk} index does this.

Before the team can arrive at C_{pk}, the values for two additional indices must be found. C_{pl} and C_{pu} are measures of single-sided specification limits, while C_{pk} is for two-sided specification limits. All three indices take into consideration just where the process average is located. C_{pk} is the measure of true process capability and is the smaller value when C_{pl} and C_{pu} are compared. The three values are found in the following manner:

Using the values from the example charts in Figure 8.7: $C_{pl} = \bar{\bar{X}}$ minus the lower specification limit divided by 3 times $\hat{\sigma}$

$$= (5.11 - 3) / 6.06$$
$$= .35$$

C_{pu} = The upper specification limit minus $\overline{\overline{X}}$ divided by 3 times $\hat{\sigma}$

$$= (7 - 5.11) / 6.06$$

$$= .31$$

C_{pk} = min(C_{pl}, C_{pu})

$$= \text{Value of the smaller}$$

$$= .31$$

When C_{pl} and C_{pu} are equal, the process is exactly centered.

A comparison of the manner in which C_p and C_{pk} interrelate in the above example shows that C_{pk} *can never be larger than C_p* in the same distribution. The further the center of the distribution moves from the center value of the specification limits, the more C_{pk} deteriorates. Therefore, a major value for computing C_p lies in seeing how much better C_{pk} could be were the center of the distribution closer to the center value of the specification limits. The value of C_{pk} is far more definitive. When C_{pk} is greater than 1, the total distribution is within the specification limits. In other words, both ends of the spread across the 6-σ width of the distribution are within specification. If C_{pk} is found to be exactly 1, then one end of the spread falls exactly on one of the specification limits. Were C_{pk} to be any value less than 1, then at least some portion of the distribution lies outside the specification limits.

In summary, C_p is a measure of the spread of a distribution across the 6-σ width of a distribution, without consideration of whether the distribution is centered. C_{pk}, on the other hand, measures the spread of a distribution and the manner in which the spread is centered. A view of C_p and C_{pk} is provided in the *DataMyte Handbook*,[10] an excellent and concise summary of both. It states that:

> C_p answers whether the distribution of individuals *could* fit within the tolerances (if centered). C_{pk} answers whether the distribution *does*.

EPILOG

The challenge for an engineering manager today is greater than at any time in the past. Modern advanced technology requires constant upgrading of individual knowledge at every level, but especially in the ranks of management. An engineer who finds himself promoted to a management position must provide the leadership required for his employees to accomplish the implementation of new and innovative designs and processes, as well as foster an atmosphere for the improvement of existing processes.

Fortunately, there is a way.

By practicing the suggestions for management presented in the early chapters of this text, an engineering manager can feel confident that his management style will improve. By implementing the step-by-step methods for process improvement found throughout the text, the engineering manager can produce a team environment within the scope of his own jurisdiction that will endow each member of the team with a feeling of problem-solution ownership. That single accomplishment is probably the most important idea presented here. Problem-solution ownership provides an immediate target for each employee involved. It supplies the employee with a dedication toward implementation of the solution that can be obtained in no other way.

The astute engineering manager *can* succeed in implementing TEQM by applying the concepts in this text; however, rest assured, it will not be easy.

We had a short saying during my time in the U.S. Navy that fits this situation well:

If you're going nowhere,
put some wind behind you,
reset your sails,
and change course.

You must continue to try. Good luck.

REFERENCES

1. McGregor, Douglas. *The Human Side of Enterprise*. New York: McGraw-Hill, 1970.
2. Herzberg, Frederick. *The Motivation to Work*. New York: John Wiley and Sons, 1979.
3. Maslow, Abraham H. *Motivation and Personality*. New York: Harper and Row, 1970.
4. Boulding, Kenneth E. General systems theory: the skeleton of science. *Management Science*, April 1966.
5. Johnson, R. A., Kast, F. E., and Rosenweig, J. *The Theory and Management of Systems*. New York: McGraw-Hill, 1967.
6. Christofono, R. Fundamental Statistical Process Control. Ron Christofono Workshop Series, 1984.
7. *Webster's Student Dictionary*. Closter, NJ: Sharen Publications, 1990.
8. Cleland, D. I., and King, W. R. *Management: A Systems Approach*. New York: McGraw-Hill, 1972.
9. Juran, J. M. *Juran's Quality Control Handbook*. New York: McGraw-Hill, 1988.

10. DataMyte Corporation. *DataMyte Handbook*. Minnetonka, MN: DataMyte Corporation, 1989.
11. Amsden, R. T., Butler, H. E., and Amsden, D. M. *SPC Simplified: Practical Steps to Quality*. White Plains, NY: UNIPUB, Kraus International Publications, 1986.
12. Hradesky, J. *Productivity and Quality Improvement*. New York: McGraw-Hill, 1988.
13. *The Memory Jogger*. Methuen, MA: GOAL/OPC, 1988.

INDEX

For Product Safety Concerns and Information please contact our EU
representative GPSR@taylorandfrancis.com
Taylor & Francis Verlag GmbH, Kaufingerstraße 24, 80331 München, Germany

www.ingramcontent.com/pod-product-compliance
Ingram Content Group UK Ltd.
Pitfield, Milton Keynes, MK11 3LW, UK
UKHW021826240425
457818UK00006B/92